THE ELECTRONIC

A C Fischer-Cripps

Institute of Physics Publishing
Bristol and Philadelphia

British Library Cataloguing-in-Publication Data

A catalogue record for this book is available from the British Library.

ISBN 0 7503 1012 X

Library of Congress Cataloging-in-Publication Data are available

Commissioning Editor: Tom Spicer
Production Editor: Simon Laurenson
Production Control: Sarah Plenty
Cover Design: Frédérique Swist
Marketing: Nicola Newey, Louise Higham and Ben Thomas

Published by Institute of Physics Publishing, wholly owned by The Institute of Physics, London

Institute of Physics Publishing, Dirac House, Temple Back, Bristol BS1 6BE, UK

US Office: Institute of Physics Publishing, The Public Ledger Building, Suite 929, 150 South Independence Mall West, Philadelphia, PA 19106, USA

Printed in the UK by MPG Books Ltd, Bodmin, Cornwall

This book is dedicated to
Robert Winston Cheary
who taught me all that I know
about electronics.

Contents

Preface

This book is designed to help you understand the basic principles of electronics. The combination of succinct, but detailed explanations, review questions, and laboratory experiments work together to provide a consistent and logical account of the way in which basic electronics circuits are designed and how they work. The book arose out of a series of lectures that I attended as a student, and then later in life, as the lecturer. I am indebted to the late Robert Cheary, who presented this course for many years at the University of Technology, Sydney. I also express my appreciation to Walter Kalceff and Les Kirkup, my co-presenters, and Anthony Wong, who assisted me in the laboratory with many generations of enthusiastic students. Finally, I thank Tom Spicer and the editorial and production team at Institute of Physics Publishing for their very professional and helpful approach to the whole publication process.

I hope that you will find this book a useful companion in your study of electronics.

Tony Fischer-Cripps,
Killarney Heights, Australia, 2004

1. Electricity

Summary

$$F = k\frac{q_1 q_2}{d^2}$$

Force between two charges where $k = \dfrac{1}{4\pi\varepsilon_o}$

$$F = q_1 E$$

Force on a charge in a field

$$E = 4\pi k\frac{Q}{A}$$

Electric field - point charge

$$\overline{E} = k\frac{q\hat{r}}{r^2}$$

Electric field - point charge

$$\phi = EA$$

Electric flux

$$I = A\big(q_1 n_1 v_1 + (-q_2)n_2(-v_2)\big)$$

$$i = \frac{dq}{dt}$$

Electric current

$$\frac{W}{q} = Ed$$

Electric potential

$$\frac{V}{I} = R$$

Ohm's law

$$P = VI = I^2 R$$

Power - resistor

$$R = \rho\frac{l}{A}$$

Resistivity

$$C = \frac{Q}{V} = \varepsilon_o\frac{A}{d}$$

Capacitance

$$L = \mu_o A\frac{N^2}{l}$$

Inductance

$$U = \frac{1}{2}CV^2$$

Energy - capacitor

$$U = \frac{1}{2}LI^2$$

Energy - inductor

$$R_{AB} = R_1 + R_2$$

Resistors - series

$$\frac{1}{R_{AB}} = \frac{1}{R_1} + \frac{1}{R_2}$$

Resistors - parallel

$$R_{AB} = \frac{R_1 R_2}{R_1 + R_2}$$

1.1 Electricity

Consider a circuit in which a battery is connected to a light bulb through a switch.

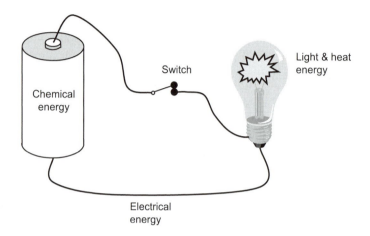

In this simple electrical system, **chemical energy** is converted into **electrical energy** in the battery. The electrical energy travels along the wires to the light bulb where it is converted into heat and light. The switch is used to interrupt the flow of electrical energy to the light bulb.

Although such an electrical system may seem commonplace to us now, it was only invented about 100 years ago. For thousands of years before this, light and heat were obtained by burning oil or some other combustible fuel (e.g. wood). Although the concept of electric charge was known to the ancient Greeks, and electricity as we know it was well-studied in the 19th century, it remained a scientific curiosity for many years until it was put to use in an engineering sense.

In the early part of the 20th century, electrical engineering was concerned with motors, generators and generally large scale electrical machines. In the second half of the 20th century, advances in the understanding of the electronic structure of matter lead to the emergence of the new field of electronics. Initially, electronic circuits were built around relatively large scale devices such as thermionic valves. Later, the functionality of valves was implemented using solid-state components through the use of semiconductors.

1.2 Electric charge

Electrical (and magnetic) effects are a consequence of a property of matter called **electric charge**. Experiments show that there are two types of charge that we label **positive** and **negative**. Experiments also show that unlike charges attract and like charges repel.

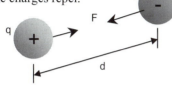

The charge on a body usually refers to its *excess* or net charge. The smallest unit of charge is that on one electron $e = -1.60219 \times 10^{-19}$ coulombs.

The force of attraction or repulsion can be calculated using **Coulomb's law:**

- if F is positive, the charges **repel** each other;
- if F is negative, the charges **attract** each other.

$$F = k\frac{q_1 q_2}{d^2}$$

Magnitude of the charges

Constant 9×10^9 Nm^2C^{-2}

Distance between charges

$$k = \frac{1}{4\pi\varepsilon_0}$$

→ the **permittivity of free space** $= 8.85 \times 10^{-12}$ farads/metre

If the two charges are in some substance, eg. air, then the **coulomb force** is reduced. Instead of using ε_0, we must use ε for the substance. Often, the **relative permittivity** ε_r is specified.

$$\varepsilon_r = \frac{\varepsilon}{\varepsilon_0}$$

Material	ε_r
Vacuum	1
Water	80
Glass	8

Now,

1. imagine that one of the charges is hidden from view;

2. the other charge still experiences the coulomb force and thus we say it is acted upon by an **electric field**;

3. if a **test charge** experiences a force when placed in a certain place, then an electric field exists at that place. The direction of the field is taken to be that in which a positive test charge would move in the field.

q_1 F q_2 E

$$F = k\frac{q_1 q_2}{d^2}$$

$$\text{let } E = k\frac{q_2}{d^2}$$

$$\text{thus } F = q_1 E$$

Note: the origin of the field E may be due to the presence of many charges but the magnitude and direction of the resultant field E can be obtained by measuring the force F on a single test charge q.

1.3 Electric flux

An electric field may be represented by lines of force. The total number of lines is called the **electric flux**. The number of lines per unit cross-sectional area is the **electric field intensity**, or simply, the magnitude of the electric field.

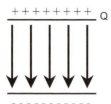

Uniform electric field between two charged parallel plates

$$E = 4\pi k \frac{Q}{A}$$

Non-uniform field surrounding a point charge

$$E = k \frac{q}{r^2}$$

• Arrows point in direction of path taken by a positive test charge placed in the field.
• Number density of lines crossing an area A indicates electric field intensity.
• Lines of force start from a positive charge and always terminate on a negative charge (even for an **isolated charge** where the corresponding negative charge may be quite some distance away).

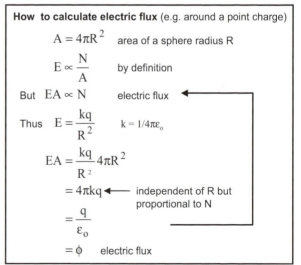

Note: for an isolated charge (or charged object) the termination charge is so far away that it contributes little to the field. When the two charges are close together, such as in the parallel plates, both positive and negative charges contribute to the strength of the field. For the plates, Q is the charge on either plate, a factor of 2 has already been included in the formula.

How to calculate electric flux (e.g. around a point charge)

$$A = 4\pi R^2 \quad \text{area of a sphere radius R}$$

$$E \propto \frac{N}{A} \quad \text{by definition}$$

But $EA \propto N$ electric flux

Thus $E = \dfrac{kq}{R^2}$ $k = 1/4\pi\varepsilon_o$

$$EA = \frac{kq}{R^2} 4\pi R^2$$

$$= 4\pi kq \longleftarrow \text{independent of R but proportional to N}$$

$$= \frac{q}{\varepsilon_o}$$

$$= \phi \quad \text{electric flux}$$

1.4 Conductors and insulators

Atoms consist of a positively charged **nucleus** surrounded by negatively charged **electrons**. Solids consist of a fixed arrangement of atoms usually arranged in a lattice. The position of individual atoms within a solid remains constant because **chemical bonds** hold the atoms in place. The behaviour of the outer electrons of atoms is responsible for the formation of chemical bonds. These outer shell electrons are called **valence electrons**.

Valence electrons

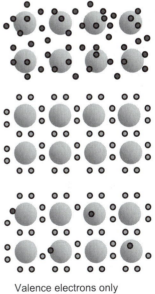

1. **Conductors**

   Valence electrons are weakly bound to the atomic lattice and are free to move about from atom to atom.

2. **Insulators**

   Valence electrons are tightly bound to the atomic lattice and are fixed in position.

3. **Semiconductors**

   In **semiconductors**, valence electrons within the crystal structure of the material are not as strongly bound to the atomic lattice and if given enough energy, may become mobile and free to move just like in a conductor.

Valence electrons only shown in these figures

Electrons, especially in conductors, are **mobile charge carriers** (they have a charge, and they are mobile within the atomic lattice).

1.5 Electric current

Mobile charge carriers may be either positively charged (e.g. positive ions in solution) or negatively charged (e.g. negative ions, loosely bound valence electrons). Consider the movement during a time Δt of positive and negative charge carriers in a **conductor** of cross-sectional area A and length l placed in an electric field E:

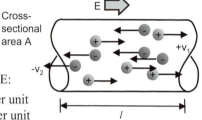

Cross-sectional area A

Let there be n_1 positive carriers per unit volume and n_2 negative carriers per unit volume. Charge carriers move with drift velocities v_1 and $-v_2$. In time Δt, each particle moves a distance $l = v_1\Delta t$ and $l = v_2\Delta t$.

The total positive charge exiting from the right (and entering from the left) during Δt is thus:

$$Q^+ = q_1 n_1 (v_1\Delta t)A$$

Total charge coulombs

Charge on one mobile carrier

No. of charge carriers per unit volume

Volume

The total negative charge exiting from the left is $Q^- = n_2(-q_2)(-v_2\Delta t)A$. The total net movement of charge during Δt is thus:

$$Q^+ + Q^- = q_1 n_1 v_1 A\Delta t + (-q_2)n_2(-v_2)A\Delta t$$

The total charge passing any given point in coulombs per second is called electric current:

$$\frac{Q^+ + Q^-}{\Delta t} = q_1 n_1 v_1 A + (-q_2)n_2(-v_2)A$$

1 amp is the rate of flow of electric charge when one coulomb of electric charge passes a given point in an electric circuit in one second.

$$I = A\big(q_1 n_1 v_1 + (-q_2)n_2(-v_2)\big)$$

Current density $J = I/A$

amps m^{-2} or coulombs m^{-2} s^{-1}

In general,

$$i = \frac{dq}{dt}$$

Lower case quantities refer to instantaneous values. Upper case refers to steady-state or DC values.

In metallic conductors, the mobile charge carriers are negatively charged electrons hence $n_1 = 0$.

Note that the **amp** is a measure of quantity of charge per second and provides no information about the net drift velocity of the charge carriers (≈ 0.1 mm s^{-1}).

1.6 Conventional current

Electric current involves the net flow of **electrical charge carriers,** which, in a metallic conductor, are negatively charged electrons. Often in circuit analysis, the physical nature of the actual flow of charge is not important - whether it be the flow of free electrons or the movement of positive ions in a solution.

This happens due to the movement of both positive and negative ions within the solution inside the cell. ⟶

But, in the 1830s, no one had heard of the **electron**. At that time, Faraday noticed that when current flowed through a wire connected to a chemical cell, the **anode** (positive) lost weight and the **cathode** (negative) gained weight, hence it was concluded that charge carriers flowed through the wire from positive to negative.

We now know that positive ions in the solution move towards the anode and negative ions towards the cathode. Electrons in the wire go from the cathode (−) to the anode (+).

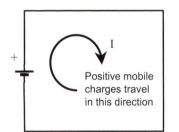

Positive mobile charges travel in this direction

There are two types of charge carriers: positive and negative. For historical reasons, all laws and rules for electric circuits are based on the direction that would be taken by positive charge carriers. Thus, in all circuit analysis, *imagine* that current flows due to the motion of positive charge carriers. Current then travels from positive to negative. This is called **conventional current**.

If we need to refer to the actual *physical process* of conduction, then we refer to the specific charge carriers appropriate to the conductor being considered.

1.7 Potential difference

When a charge carrier is moved from one point to another in an electric field, its potential energy is changed since this movement involved a force F moving through a distance.

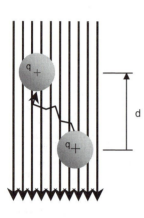

If q+ is moved *against* the field, then work is done *on* the charge and the **potential energy** of the charge is increased.

$$F = qE$$

The work done is thus:

$$W = Fd$$

$$= qEd \quad \text{joules}$$

The work per unit charge is called the **electrical potential**:

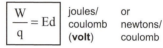

$\dfrac{W}{q} = Ed$	joules/ coulomb (**volt**)	or newtons/ coulomb

If a charged particle is released in the field, then work is done on the particle and it acquires kinetic energy. The force acting on the particle is proportional to the field strength E. The stronger the field, the larger the force – the greater the acceleration and the greater the rate at which the charged particle acquires **kinetic energy**.

A uniform electric field E exists between two parallel charged plates since a positive test charge placed anywhere within this region will experience a downwards force of uniform value. The electric field also represents a **potential gradient**.

If the negative side of the circuit is grounded, then the electrical potential at the negative plate is zero and increases uniformly through the space between the plates to the top plate where it is +V.

The potential gradient (in volts per metre) is numerically equal to the **electric field strength** (newtons per coulomb) but is opposite in direction.

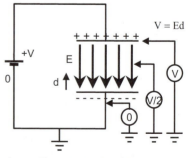

In a uniform electric field, the potential decreases uniformly along the field lines and is a potential gradient.

1.8 Resistance

A voltage source, utilising chemical or mechanical means, raises the electrical potential of mobile charge carriers (usually electrons) within it. There is a net build up of charge at the terminals of a voltage source. This net charge results in an **electric field** which is channelled through the conductor. Mobile electrons within the conductor thus experience an electric force and are accelerated.

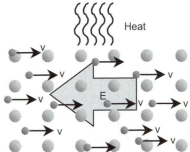

However, as soon as these electrons move through the conductor, they suffer collisions with other electrons and fixed atoms and lose velocity and thus some of their kinetic energy. Some of the fixed atoms correspondingly acquire **internal energy** (vibrational motion) and the **temperature** of the conductor rises. After collision, electrons are accelerated once more and again suffer more collisions.

Note: negatively charged electrons move in opposite direction to that of electric field.

Alternate accelerations and decelerations result in a net average velocity of the mobile electrons (called the **drift velocity**) which constitutes an electric current. **Electrical potential energy** is converted into **heat** within the **conductor**. The decelerations arising from collisions constitute electrical **resistance**.

Experiments show that, for a particular specimen of material, when the applied voltage is increased, the current increases. For most materials, doubling the voltage results in a doubling of the current. That is, the current is directly proportional to the current: $I \propto V$

The constant of proportionality is called the **resistance**. Resistance limits the current flow through a material for a particular applied voltage.

$$\frac{V}{I} = R \quad \textbf{Ohm's law}$$

Units: **ohms** Ω

The rate at which electrical potential energy is converted into heat is the **power** dissipated by the resistor. Since electrical potential is joules/coulomb, and current is measured in coulombs/second, then the product of voltage and current gives joules/second which is **power** (in **watts**).

$$P = VI$$
but $\quad V = IR$
thus $\quad \boxed{P = I^2 R}$

1.9 Resistivity

Experiments show that the **resistance** of a particular specimen of material (at a constant temperature) depends on three things:

- the length of the conductor, l
- the cross-sectional area of the conductor, A
- the type of material, ρ

The material property which characterises the ability for a particular material to conduct electricity is called the **resistivity** ρ (the inverse of which is the **conductivity** σ).

Material	ρ Ωm @ 20 °C
Silver	1.64×10^{-8}
Copper	1.72×10^{-8}
Aluminium	2.83×10^{-8}
Tungsten	5.5×10^{-8}

The resistance R (in ohms) of a particular length l of material of cross-sectional area A is given by:

Now, $V = IR$

$$R = \rho \frac{l}{A}$$

The units of ρ are Ωm, the units of σ are Sm^{-1}.

hence $= I \dfrac{\rho l}{A}$

$\dfrac{V}{l} = \rho \dfrac{I}{A}$

The quantity I/A is called the **current density** J.

but $E = \dfrac{V}{l}$

thus $E = \rho \dfrac{I}{A}$

The **number density** of **mobile charge carriers** n depends on the material. If the number density is large, then, if E (and hence v) is held constant , the **resistivity** must be small. Thus, the resistivity depends inversely on n. **Insulators** have a high resistivity since n is very small. **Conductors** have a low resistivity because n is very large.

$= \rho \dfrac{\sum nqvA}{A}$

Sum of the positive and negative mobile charge carriers $\Sigma nqvA = n_1 q_1 v_1 + n_2 (-q_2)(-v_2)A$

For a metal, only one type of mobile charge carrier.

$$\rho = \frac{E}{nqv}$$

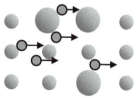

For a particular specimen of material, n, q, A and l are a constant. Increasing the applied field E results in an increase in the **drift velocity** v and hence an increase in current I.

The **resistivity** of a pure substance is lower than that of one containing impurities because the mobile electrons are more likely to travel further and acquire a larger velocity when there is a regular array of stationary atoms in the conductor.

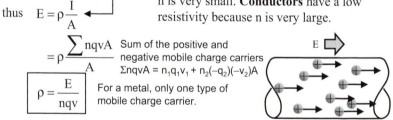

Presence of impurity atoms decreases the average drift velocity.

1.10 Variation of resistance

Consider an applied voltage
which generates an electric field
E within a conductor of resistance
R and of length *l* and area A.

1. Variation with area

Evidently, if the area A is increased, there will be more mobile charge
carriers available to move past a given point during a time Δt under the
influence of the field and the current I increases. Thus, for a particular
specimen, the resistance decreases with increasing cross-sectional area.

2. Variation with length

Now, the field E acts over a length *l*.

$$V = El$$

If the applied voltage is kept
constant, then it is evident that if *l* is
increased, E must decrease. The
drift velocity depends on E so that if
E decreases, then so does v, and
hence so does the current.

3. Variation with temperature

Increasing the **temperature** increases
the random thermal motion of the atoms in the conductor thus increasing
the chance of collision with a mobile electron thus reducing the average
drift velocity and increasing the resistivity. Different materials respond to
temperature according to the **temperature coefficient of resistivity** α.

$$\rho_T = \rho_0[1 + \alpha(T - T_0)]$$
$$R_T = R_0[1 + \alpha(T - T_0)]$$

R_T = resistance at T
R_0 = resistance at T_0 (usually 0 °C)

This formula applies to
a conductor, not a
semi-conductor.

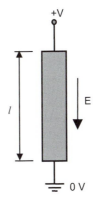

Material	α (C^{-1}Ω^{-1})
Tungsten	4.5×10^{-3}
Platinum	3.0×10^{-3}
Copper	3.9×10^{-3}

1.11 Resistor circuits

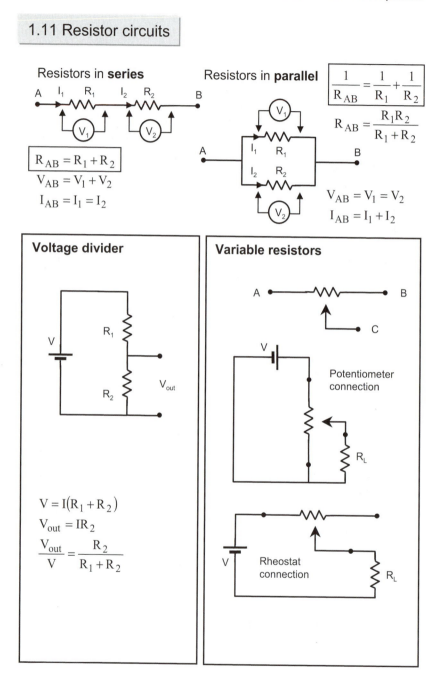

Resistors in **series**

A I_1 R_1 I_2 R_2 B

V_1 V_2

$$R_{AB} = R_1 + R_2$$
$$V_{AB} = V_1 + V_2$$
$$I_{AB} = I_1 = I_2$$

Resistors in **parallel**

$$\frac{1}{R_{AB}} = \frac{1}{R_1} + \frac{1}{R_2}$$

$$R_{AB} = \frac{R_1 R_2}{R_1 + R_2}$$

V_1

A I_1 R_1 B

I_2 R_2

V_2

$$V_{AB} = V_1 = V_2$$
$$I_{AB} = I_1 + I_2$$

Voltage divider

V

R_1

R_2 V_{out}

$$V = I(R_1 + R_2)$$
$$V_{out} = IR_2$$
$$\frac{V_{out}}{V} = \frac{R_2}{R_1 + R_2}$$

Variable resistors

A ——— B

C

V

Potentiometer
connection

R_L

V

Rheostat
connection

R_L

1.12 Emf

Consider a **chemical cell**:

Excess of positive charge
+

Electric field
E →

Excess of negative charge
−

┌ rather than "electrostatic"

(a) (b)

Chemical attractions cause positive charges to build up at (a) and negative charges at (b).

Electrostatic repulsion due to build-up of positive charge at (a) eventually becomes equal to the chemical attractions tending to deposit more positive charges and system reaches equilibrium.

The term **emf** (electromotive force) is defined as the amount of energy expended by the cell in moving 1 coulomb of charge from (b) to (a) *within* the cell.

At **open circuit**, emf $= V_{ab}$

"force" is poor choice of words since emf is really "energy" (joules per coulomb)

Now connect an external load R_L across (a) and (b). The **terminal voltage** V_{ab} is now reduced.

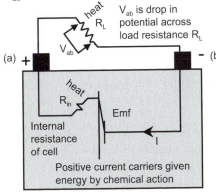

heat
R_L
V_{ab}

V_{ab} is drop in potential across load resistance R_L

(a) +

− (b)

heat
R_{in}

Internal resistance of cell

Emf

I

Positive current carriers given energy by chemical action

Loss of positive charge from (a) reduces the accumulated charge at (a) and hence chemical reactions proceed and more positive charges are shifted from (b) to (a) within the cell to make up for those leaving through the external circuit. Thus, there is a steady flow of positive charge through the cell and through the wire.

Assume positive carriers – **conventional current** flow.

The circuit has been drawn to emphasize where potential drops and rises occur.

But, the continuous conversion of chemical potential energy to electrical energy is not 100% efficient. Charge moving within the cell encounters **internal resistance** which, in the presence of a current I, means a voltage drop so that:

At "closed circuit", emf $= V_{ab} + IR_{in}$

1.13 Capacitance

Consider two **parallel plates** across which is placed a voltage V.

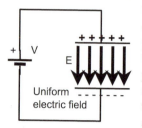

When a voltage V is connected across the plates, current begins to flow as charge builds up on each plate. In the diagram, **negative charge** builds up on the lower plate and **positive charge** on the upper plate. The accumulated charge on the two plates establishes an electric field between them. Since there is an electric field between the plates, there is an **electrical potential difference** between them.

For a **point charge** in space, E depends on the distance away from the charge: $E = \dfrac{1}{4\pi\varepsilon_o}\dfrac{q}{d^2}$

But, for **parallel plates** holding a total charge Q on each plate, calculations show that the electric field E in the region between the plates is proportional to the magnitude of the charge Q and inversely proportional to the area A of the plates. For a given accumulated charge +Q and −Q on each plate, the field E is independent of the distance between the plates.

Permittivity of free space = 8.85×10^{-12} farads m^{-1}

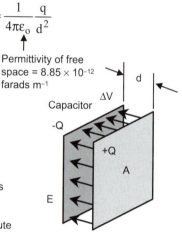

Capacitor

$$E = \frac{Q}{\varepsilon_0 A}$$

Now, $V = Ed$

thus $V = \dfrac{Q}{\varepsilon_0 A}d$

Q in these formulas refers to the charge on ONE plate. Both positive and negative charges contribute to the field E. A factor of 2 has already been included in these formulas.

A charged particle released between the plates will experience an accelerating force.

Capacitance is defined as the ratio of the magnitude of the charge on each plate (+Q or −Q) to the potential difference between them.

A large capacitor will store more charge for every volt across it than a small capacitor.

$$C = \frac{Q}{V}$$ but $V = \dfrac{Q}{\varepsilon_0 A}d$

$$= Q\frac{\varepsilon_0 A}{Qd}$$

$$C = \varepsilon_0 \frac{A}{d}$$

Units: **farads**

If the space between the plates is filled with a **dielectric**, then capacitance is increased by a factor ε_r. A dielectric is an insulator whose atoms become polarised in the electric field. This adds to the **storage capacity** of the capacitor.

1.14 Capacitors

If a capacitor is charged and the voltage source V is then disconnected from it, the accumulated charge remains on the plates of the capacitor.

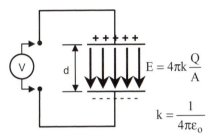

$$E = 4\pi k \frac{Q}{A}$$

$$k = \frac{1}{4\pi\varepsilon_o}$$

Since the charges on each plate are opposite, there is an **electrostatic force of attraction** between them but the charges are kept apart by the gap between the plates. In this condition, the capacitor is said to be **charged**. A **voltmeter** placed across the terminals would read the voltage V used to charge the capacitor.

When a **dielectric** is inserted in a capacitor, the molecules of the dielectric align themselves with the applied field. This alignment causes a field of opposite sign to exist within the material thus reducing the overall net field. For a given applied voltage, the total net field within the material is small for a material with a high **permittivity** ε. The permittivity is thus a measure of how easily the charges within a material line up in the presence of an applied external field.

If the plates are separated by a dielectric, then the field E is reduced. Instead of using ε_o, we must use ε for the substance. Often, the **relative permittivity** ε_r is specified.

In a conductor, charge carriers not only align but actually move under the influence of an applied field. This movement of charge carriers completely cancels the external field all together. The net electric field within a conductor placed in an external electric field is zero!

$$\varepsilon_r = \frac{\varepsilon}{\varepsilon_o}$$

Material	ε_r
Vacuum	1
Water	80
Glass	8

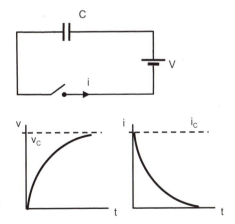

When a capacitor is connected across a voltage source, the current in the circuit is initially very large and then decreases as the capacitor charges. The voltage across the capacitor is initially zero and then rises as the capacitor charges.

1.15 Energy stored in a capacitor

Energy is required to charge a capacitor. When a capacitor is connected across a voltage source, the current in the circuit is initially very large and then decreases as the capacitor charges. The voltage across the capacitor is initially zero and then rises as the capacitor charges.

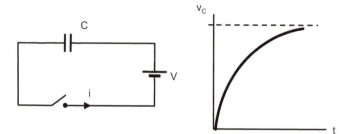

Energy is expended by the voltage source as it forces charge onto the plates of the capacitor. When fully charged, and disconnected from the voltage source, the voltage across the capacitor remains. The stored electric potential energy within the charged capacitor may be released when desired by discharging the capacitor.

Power $P = vi$ Lower case letters refer to
                       instantaneous quantities.

$$i = \frac{dq}{dt}$$

$$Pdt = vdq = dU$$

$$U = \int_0^Q vdq \qquad \text{Energy}$$

$$= \int_0^Q \frac{q}{C}dq \qquad C = \frac{q}{v}$$

$$= \frac{1}{2}\frac{Q^2}{C} \qquad C = \frac{Q}{V}$$

Energy stored $\boxed{U = \frac{1}{2}CV^2}$
in a capacitor

1.16 Capacitor circuits

Capacitors in **series**

$$Q = Q_1 = Q_2 = Q_3$$
$$V = V_1 + V_2 + V_3$$
$$C_{total} = \frac{Q}{V}$$
$$V_1 = \frac{Q}{C_1}; V_2 = \frac{Q}{C_2}; V_3 = \frac{Q}{C_3} \quad \leftarrow \text{ charge on one plate}$$
$$V = Q\left(\frac{1}{C_1} + \frac{1}{C_2} + \frac{1}{C_3}\right)$$
$$\frac{1}{C_{total}} = \frac{1}{C_1} + \frac{1}{C_2} + \frac{1}{C_3}$$

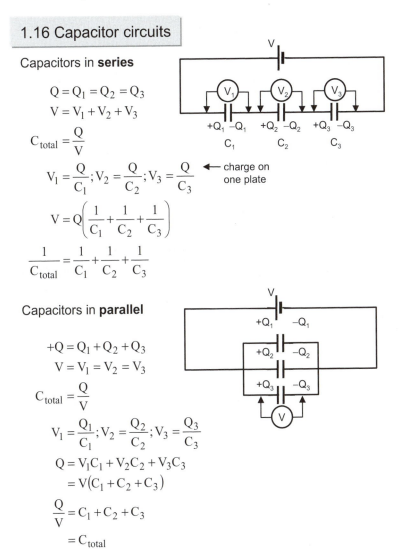

Capacitors in **parallel**

$$+Q = Q_1 + Q_2 + Q_3$$
$$V = V_1 = V_2 = V_3$$
$$C_{total} = \frac{Q}{V}$$
$$V_1 = \frac{Q_1}{C_1}; V_2 = \frac{Q_2}{C_2}; V_3 = \frac{Q_3}{C_3}$$
$$Q = V_1C_1 + V_2C_2 + V_3C_3$$
$$= V(C_1 + C_2 + C_3)$$
$$\frac{Q}{V} = C_1 + C_2 + C_3$$
$$= C_{total}$$

1.17 Inductance

In a conductor carrying a steady electric current, there is a magnetic field around the conductor. The magnetic field of a current-carrying conductor may be concentrated by winding the conductor around a tube to form a **solenoid**.

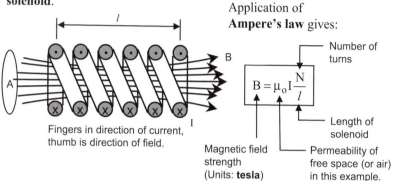

Application of **Ampere's law** gives:

$$B = \mu_o I \frac{N}{l}$$

Number of turns

Length of solenoid

Magnetic field strength (Units: **tesla**)

Permeability of free space (or air) in this example.

Fingers in direction of current, thumb is direction of field.

When the current in the coil changes, the resulting change in **magnetic field** induces an emf in the coil (**Faraday's law**).

$$V = -\frac{d\Phi}{dt}$$

$$= -\mu_o A \frac{N}{l} \frac{di}{dt} \quad \text{where} \quad B = \frac{\Phi}{A}$$

Magnetic flux

Magnetic field

Cross-sectional area of coil

But, this is the voltage induced in *each loop* of the coil. Each loop lies within a field B and experiences the changing current. The *total* voltage induced between the two ends of the coil is thus N times this:

$$V_{total} = -\mu_o A \frac{N^2}{l} \frac{di}{dt}$$

The induced voltage tends to oppose the change in current (**Lenz's law**).

Inductance: $\boxed{L = \mu_o A \frac{N^2}{l}}$ determines what voltage is induced within the coil for a given rate of change of current.

(Units: **henrys**)

1.18 Inductors

In a circuit with an inductor, when the **switch** closes, a changing current results in a changing magnetic field around the coil. This *changing* magnetic field induces a voltage (emf) in the loops (Faraday's law) which tends to oppose the applied voltage (Lenz's law). Because of the self induced opposing emf, the current in the circuit does not rise to its final value at the instant the circuit is closed, but grows at a rate which depends on the **inductance** (in **henrys**, L) and resistance (R) of the circuit. As the current increases, the *rate of change* of current decreases and the magnitude of the opposing voltage decreases. The current reaches a maximum value I when the opposing voltage drops to zero and all the voltage appears across the resistance R.

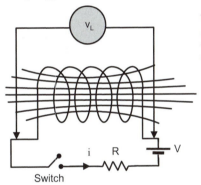

Switch

When the switch is closed, the rate of change of current is controlled by the value of L and R. Calculations show that the voltage across the inductor is given by:

$$v_L = Ve^{-\frac{Rt}{L}}$$

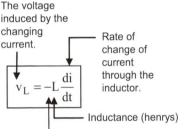

The voltage induced by the changing current.

Rate of change of current through the inductor.

$$v_L = -L\frac{di}{dt}$$

Inductance (henrys)

The minus sign indicates that if the current is decreasing (di/dt is negative) then the voltage v_L induced in the coil is positive (i.e. same direction as V). If we consider v_L as the voltage drop across the coil (in much the same way as we talk about a voltage drop across a resistor) then $v_L = Ldi/dt$. The sign depends upon your point of view: whether the inductor is considered a voltage "source" or a voltage drop.

Switch is closed

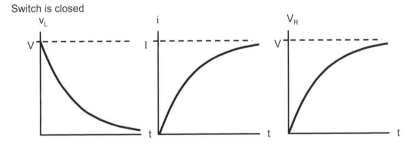

1.19 Discharge and stored energy

Establishing a current in an inductor requires energy which is stored
in the **magnetic field**. When an inductor is discharged, this energy is
released. If the inductor is discharged through a resistor as shown,
then the rate of decrease in current in the circuit is given by:

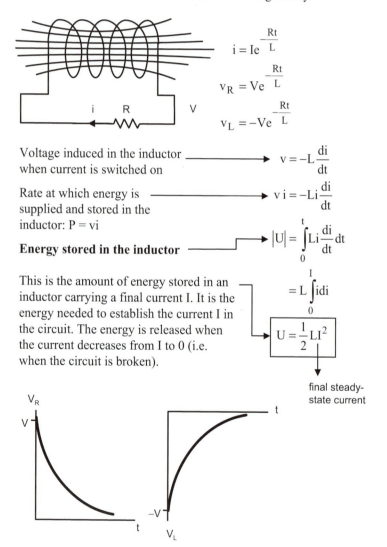

$$i = Ie^{-\frac{Rt}{L}}$$

$$v_R = Ve^{-\frac{Rt}{L}}$$

$$v_L = -Ve^{-\frac{Rt}{L}}$$

Voltage induced in the inductor
when current is switched on $\longrightarrow$ $v = -L\dfrac{di}{dt}$

Rate at which energy is
supplied and stored in the
inductor: P = vi $\longrightarrow$ $vi = -Li\dfrac{di}{dt}$

Energy stored in the inductor $\longrightarrow$ $|U| = \displaystyle\int_0^t Li\dfrac{di}{dt}dt$

This is the amount of energy stored in an
inductor carrying a final current I. It is the
energy needed to establish the current I in
the circuit. The energy is released when
the current decreases from I to 0 (i.e.
when the circuit is broken).

$$= L\int_0^I i\,di$$

$$\boxed{U = \frac{1}{2}LI^2}$$

final steady-
state current

Review questions

1. A negative charge of -0.1 μC exerts an attractive force of 0.5 N on an unknown charge at a distance of 0.5 m. Determine the magnitude (and sign) of the unknown charge.

   (Ans: 138.9 μC)

2. Three electrically charged billiard balls are placed at positions as shown in the diagram. Determine the magnitude and direction of the resulting forces on the black ball.

   $C_1 = 3$; $C_2 = -4$; $C_3 = -2$ μC
   $d_1 = 10$; $d_2 = 5$; $d_3 = 5$ cm

   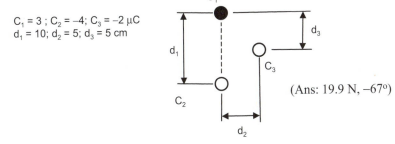

   (Ans: 19.9 N, $-67°$)

3. A uniform electric field exists between two oppositely charged parallel plates. An electron is released from rest at the surface of the negatively charged plate and strikes the surface of the positively charged plate. The distance between the plates is 2 cm and it takes 1.5×10^{-8} secs for the electron to travel from one plate to the other. (a) Find the electric field strength, and (b) the velocity of the electron as it strikes the positively charged plate.

   $m_e = 9.11 \times 10^{-31}$ kg
   $q_e = 1.6 \times 10^{-19}$ C

   (Ans: 1012 NC^{-1}, 2.67×10^6 ms^{-1})

4. A steady current of 6 A is maintained in a metallic conductor. What charge (in coulombs) is transferred through it in 1 minute?

   (Ans: 360 C)

5. Two parallel plates are 1 cm apart and are connected to a 500 V source. What force will be exerted on a single electron half way between the plates?

   (Ans: 8×10^{-15} N)

6. Calculate the resistance of 100 m of copper wire which has a diameter of 0.6 mm.

(Ans: 6.08 Ω)

7. A steady voltage source $V = 1$ V is connected to a coil which consists of a 50 m length of copper wire of radius 0.01 mm. Given that the number density of mobile electrons in copper is 8.5×10^{22} cm^{-3} and $\rho = 1.72 \times 10^{-8}$ Ω m, calculate:

(a) the electric field E which acts on the mobile electrons in the coil;
(b) the drift velocity v of the mobile electrons;
(c) the resistance R of the coil;
(d) the steady DC current I in the coil.

(Ans: 0.02 Vm^{-1}, 8.6×10^{-5} ms^{-1}, 2738 Ω, 0.36 mA)

8. A fuse in a motor vehicle electrical system has a resistance of 0.05 Ω. It is designed to blow when the power dissipation exceeds 50 W. What is the current rating of the fuse?

(Ans: 31.6 A)

9. A 12 volt battery in a motor vehicle is capable of supplying the starter motor with 150 A. It is noticed that the terminal voltage of the battery drops to 10 V when the engine is cranked over with the starter motor. Determine the internal resistance of the 12 volt battery.

(Ans: 0.013 Ω)

10. A capacitor consists of two parallel plates each of area 200 cm^2 separated by an air gap of 0.4 mm thickness. 500 V is applied. Calculate:
(a) capacitance of this capacitor;
(b) charge on each plate;
(c) energy stored in the capacitor;
(d) electric field strength between the plates.

(Ans: 442 pF, 2.24 × 10^{-7} C, 5.6 × 10^{-5} J, 1.25 × 10^6 V m^{-1})

11. A coil has an inductance of 5 H and a resistance of 20 Ω. If a DC voltage of 100 V is applied to the coil, find the energy stored in the coil when a steady maximum current has been reached.

(Ans: 62.5 J)

2. DC Circuits

Summary

Kirchhoff's laws

    1st law: Current into a junction = current out of a junction.

    2nd law: In any loop in a circuit, the sum of the voltage drops
                 equals the sum of the emf's.

Thevenin's theorem

$$R_{int} = \frac{V_{open\text{-}circuit}}{I_{short\text{-}circuit}}$$

2.1 Superposition

In a circuit containing several sources of emf, the current flowing in any branch of the circuit will be equal to the sum of the **current components** that would flow in the branch if each source of emf were to be acting alone.

Example:
Find the current in the 20 Ω
resistor in the circuit shown.

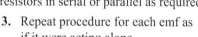

Solution:
We proceed as follows: replace
all sources with their **internal**
resistances except one and calculate the current component flowing in the
branch of interest by combining resistors in serial or parallel as required.

1. Combine 60 and 20 Ω into one resistance.

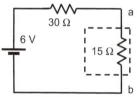

2. Analyse for unknown current *component*.

$$R_{60,20} = \left(\frac{1}{20} + \frac{1}{60}\right)^{-1}$$
$$= 15\,\Omega$$
$$R_T = 15 + 30$$
$$= 45$$
$$I_1 = \frac{6}{45}$$
$$= 0.133\,\text{A}$$
$$V_{ab} = (0.133)(15)$$
$$= 1.995\,\text{V}$$
$$I_2 = \frac{1.995}{20}$$
$$= 0.1\,\text{A} \downarrow$$

3. Repeat procedure for each emf as if it were acting alone.

$$R_{60,30} = 20\,\Omega$$
$$R_T = 20 + 20$$
$$= 40\,\Omega$$
$$I_2 = \frac{3}{40}$$
$$= 0.075\,\text{A}$$
$$R_{30,20} = 12\,\Omega$$
$$R_T = 72\,\Omega$$
$$I_3 = \frac{9}{72}$$
$$= 0.125\,\text{A}$$
$$V_{ab} = 0.125(12)$$
$$= 1.5\,\text{V}$$
$$I_2 = \frac{1.5}{20}$$
$$= 0.075\,\text{A} \downarrow$$

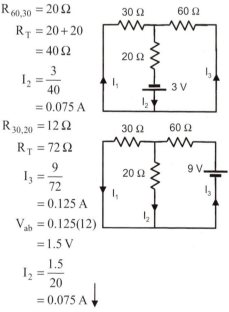

4. Add current components together for final answer.
$$I_{total} = 0.1 + 0.075 + 0.075$$
$$= 0.25\,\text{A}$$

2.2 Kirchhoff's laws

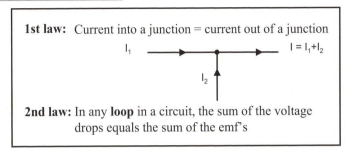

1st law: Current into a junction = current out of a junction

$I = I_1 + I_2$

2nd law: In any **loop** in a circuit, the sum of the voltage drops equals the sum of the emf's

Example:
In the circuit shown, calculate R_1, and the current through the section A-B

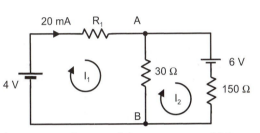

Solution:

1. Divide the circuit up into **current loops** and draw an arrow which indicates the direction of current assigned to each loop (the direction you choose need not be the correct one. If you guess wrongly, then the current will simply come out negative in the calculations).

2. Consider each loop separately:

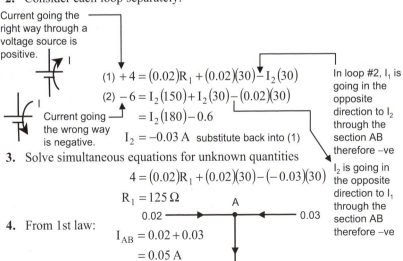

Current going the right way through a voltage source is positive.

Current going the wrong way is negative.

$(1) \quad +4 = (0.02)R_1 + (0.02)(30) - I_2(30)$

$(2) \quad -6 = I_2(150) + I_2(30) - (0.02)(30)$

$\quad\quad = I_2(180) - 0.6$

$\quad\quad I_2 = -0.03 \text{ A} \quad$ substitute back into (1)

In loop #2, I_1 is going in the opposite direction to I_2 through the section AB therefore −ve

I_2 is going in the opposite direction to I_1 through the section AB therefore −ve

3. Solve simultaneous equations for unknown quantities

$4 = (0.02)R_1 + (0.02)(30) - (-0.03)(30)$

$R_1 = 125 \,\Omega$

4. From 1st law:

$I_{AB} = 0.02 + 0.03$

$\quad\quad = 0.05 \text{ A}$

2.3 Kirchhoff's laws example

Determine the current in the two 1 Ω resistors in the following circuit.

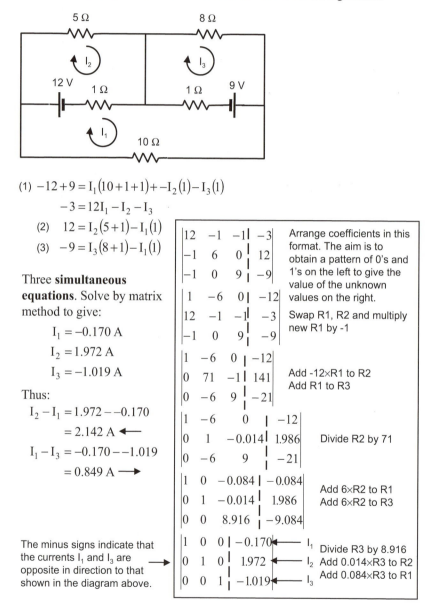

(1) $-12+9 = I_1(10+1+1)+-I_2(1)-I_3(1)$

$\qquad -3 = 12I_1 - I_2 - I_3$

(2) $12 = I_2(5+1)-I_1(1)$

(3) $-9 = I_3(8+1)-I_1(1)$

Three **simultaneous equations.** Solve by matrix method to give:

$\qquad I_1 = -0.170$ A

$\qquad I_2 = 1.972$ A

$\qquad I_3 = -1.019$ A

Thus:

$\qquad I_2 - I_1 = 1.972 - -0.170$

$\qquad\qquad = 2.142$ A ⟵

$\qquad I_1 - I_3 = -0.170 - -1.019$

$\qquad\qquad = 0.849$ A ⟶

The minus signs indicate that the currents I_1 and I_3 are opposite in direction to that shown in the diagram above.

$$\begin{vmatrix} 12 & -1 & -1 & -3 \\ -1 & 6 & 0 & 12 \\ -1 & 0 & 9 & -9 \end{vmatrix}$$
Arrange coefficients in this format. The aim is to obtain a pattern of 0's and 1's on the left to give the value of the unknown values on the right.

$$\begin{vmatrix} 1 & -6 & 0 & -12 \\ 12 & -1 & -1 & -3 \\ -1 & 0 & 9 & -9 \end{vmatrix}$$
Swap R1, R2 and multiply new R1 by -1

$$\begin{vmatrix} 1 & -6 & 0 & -12 \\ 0 & 71 & -1 & 141 \\ 0 & -6 & 9 & -21 \end{vmatrix}$$
Add -12×R1 to R2
Add R1 to R3

$$\begin{vmatrix} 1 & -6 & 0 & -12 \\ 0 & 1 & -0.014 & 1.986 \\ 0 & -6 & 9 & -21 \end{vmatrix}$$
Divide R2 by 71

$$\begin{vmatrix} 1 & 0 & -0.084 & -0.084 \\ 0 & 1 & -0.014 & 1.986 \\ 0 & 0 & 8.916 & -9.084 \end{vmatrix}$$
Add 6×R2 to R1
Add 6×R2 to R3

$$\begin{vmatrix} 1 & 0 & 0 & -0.170 \\ 0 & 1 & 0 & 1.972 \\ 0 & 0 & 1 & -1.019 \end{vmatrix}$$
⟵ I_1 Divide R3 by 8.916
⟵ I_2 Add 0.014×R3 to R2
⟵ I_3 Add 0.084×R3 to R1

2.4 Thevenin's theorem

Consider a voltage source V_T with **internal resistance** R_{int}

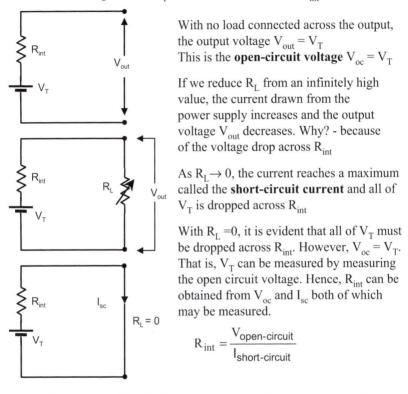

With no load connected across the output, the output voltage $V_{out} = V_T$
This is the **open-circuit voltage** $V_{oc} = V_T$

If we reduce R_L from an infinitely high value, the current drawn from the power supply increases and the output voltage V_{out} decreases. Why? - because of the voltage drop across R_{int}

As $R_L \rightarrow 0$, the current reaches a maximum called the **short-circuit current** and all of V_T is dropped across R_{int}

With $R_L = 0$, it is evident that all of V_T must be dropped across R_{int}. However, $V_{oc} = V_T$. That is, V_T can be measured by measuring the open circuit voltage. Hence, R_{int} can be obtained from V_{oc} and I_{sc} both of which may be measured.

$$R_{int} = \frac{V_{open-circuit}}{I_{short-circuit}}$$

V_T and R_{int} are useful tools for reducing a complicated power supply circuit to a simpler circuit. This is Thevenin's theorem. That is, any **two-terminal voltage source**, no matter how complicated, can be represented by V_T and R_{int}

Analysis	Measurement
1. Calculate open circuit voltage V_T using Kirchhoff or superposition 2. Determine R_{int} by replacing all internal voltage sources with their internal resistances and analysing resistance network	1. Or, measure short circuit current and open circuit voltage with multimeter to obtain R_{int}

2.5 Thevenin's theorem example

Reduce this circuit to a single voltage source V_T and internal resistance R_{int}

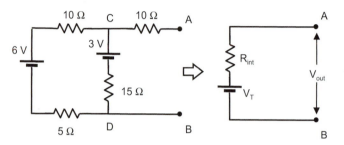

(1) Determine V_T by calculating **open circuit voltage** using Kirchhoff or superposition

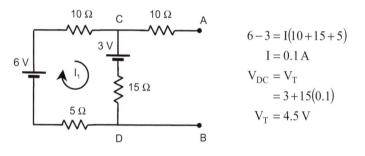

$$6 - 3 = I(10 + 15 + 5)$$
$$I = 0.1\,A$$
$$V_{DC} = V_T$$
$$= 3 + 15(0.1)$$
$$V_T = 4.5\,V$$

(2) Determine R_{int} by replacing all emf's with their **internal resistances** (zero in this example).

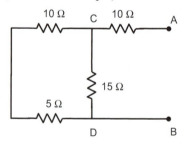

Look back into terminals A and B and calculate total resistance.

$$R_{int} = 10 + \frac{(15)(15)}{30}$$
$$= 17.5\,\Omega$$

(3) Calculate **short circuit current** if desired.

$$I_{sc} = \frac{V_{oc}}{R_{int}} = \frac{4.5}{17.5} = 257\,mA$$

2.6 Norton's theorem

Imagine a black box contains a special generator that produces a completely variable voltage but always produces a **constant current** I (equal to the **short-circuit current**).

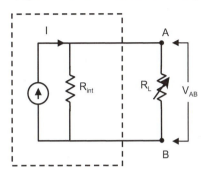

I is a constant
V_{AB} varies as R_L varies

R_{int} must be in parallel with the power source so that I remains constant when $R_L \to$ infinity. When $R_L = 0$, $I = I_{sc}$ = constant.

Consider open circuit conditions:

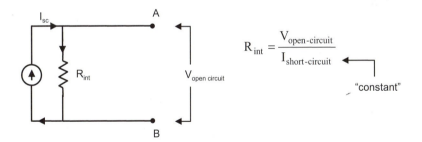

$$R_{int} = \frac{V_{open\text{-}circuit}}{I_{short\text{-}circuit}}$$

"constant"

Circuits (or parts thereof) may be reduced to **equivalent circuits** in terms of either **constant voltage** (Thevenin) or **constant current** (Norton) sources with an internal resistance. We shall see shortly that Thevenin and Norton equivalent circuits are exactly the same thing from the point of view of the voltage and current seen by an external load.

2.7 Norton's theorem example

Reduce this circuit to a constant current source I_{sc} with a parallel internal resistance R_{int}

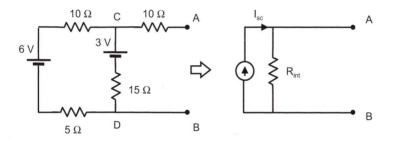

(1) Determine the open circuit voltage by Kirchhoff or superposition

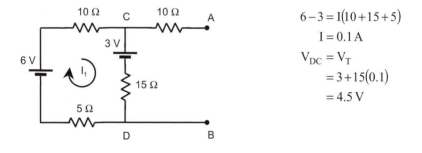

$$6 - 3 = I(10 + 15 + 5)$$
$$I = 0.1\,A$$
$$V_{DC} = V_T$$
$$= 3 + 15(0.1)$$
$$= 4.5\,V$$

(2) Determine R_{int} by replacing all emf's with their internal resistances (zero in this example).

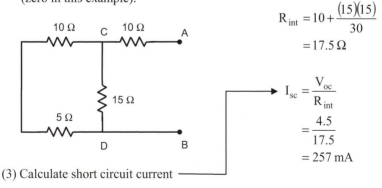

$$R_{int} = 10 + \frac{(15)(15)}{30}$$
$$= 17.5\,\Omega$$

$$I_{sc} = \frac{V_{oc}}{R_{int}}$$
$$= \frac{4.5}{17.5}$$
$$= 257\,mA$$

(3) Calculate short circuit current

2.8 Equivalence of Norton and Thevenin

Consider these two representations of a **two-terminal power supply**. If both of these can replace the actual circuitry of the power supply and appear to be the same from the point of view of whatever is connected to the output terminals A and B, then the terminal voltage V_{AB} and load current I_L must be the same for each circuit.

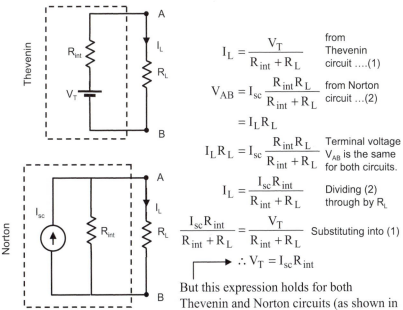

$$I_L = \frac{V_T}{R_{int} + R_L}$$ from Thevenin circuit ....(1)

$$V_{AB} = I_{sc} \frac{R_{int} R_L}{R_{int} + R_L}$$ from Norton circuit ...(2)

$$= I_L R_L$$

$$I_L R_L = I_{sc} \frac{R_{int} R_L}{R_{int} + R_L}$$ Terminal voltage V_{AB} is the same for both circuits.

$$I_L = \frac{I_{sc} R_{int}}{R_{int} + R_L}$$ Dividing (2) through by R_L

$$\frac{I_{sc} R_{int}}{R_{int} + R_L} = \frac{V_T}{R_{int} + R_L}$$ Substituting into (1)

$$\therefore V_T = I_{sc} R_{int}$$

But this expression holds for both Thevenin and Norton circuits (as shown in previous pages). Hence, V_T, I_{sc} and R_{int} are exactly the same for each circuit.

Equivalence of Norton and Thevenin

Thevenin and Norton equivalent circuits are two different ways of representing a complicated circuit, either as a simple series or a simple parallel circuit. Which one do we use? It depends on the circuit being analysed. For transistor circuits, the Norton equivalent circuit is better and leads to a simpler analysis because transistors are mostly current controlling devices.

2.9 Maximum power transfer

Transfer from an output device (e.g. amplifier) to an input device
(e.g. loudspeaker)

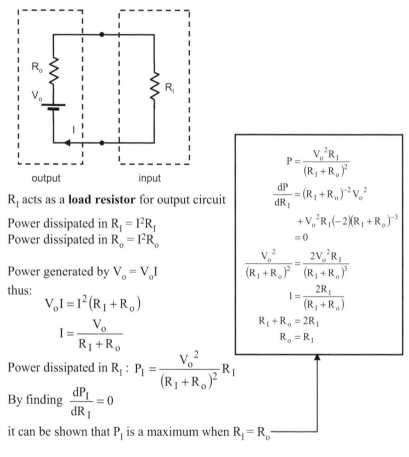

output   input

R_I acts as a **load resistor** for output circuit

Power dissipated in $R_I = I^2 R_I$
Power dissipated in $R_o = I^2 R_o$

Power generated by $V_o = V_o I$

thus:

$$V_o I = I^2 (R_I + R_o)$$

$$I = \frac{V_o}{R_I + R_o}$$

Power dissipated in R_I : $P_I = \frac{V_o^2}{(R_I + R_o)^2} R_I$

By finding $\dfrac{dP_I}{dR_I} = 0$

it can be shown that P_I is a maximum when $R_I = R_o$

$$P = \frac{V_o^2 R_I}{(R_I + R_o)^2}$$

$$\frac{dP}{dR_I} = (R_I + R_o)^{-2} V_o^2$$

$$+ V_o^2 R_I (-2)(R_I + R_o)^{-3}$$

$$= 0$$

$$\frac{V_o^2}{(R_I + R_o)^2} = \frac{2V_o^2 R_I}{(R_I + R_o)^3}$$

$$1 = \frac{2R_I}{(R_I + R_o)}$$

$$R_I + R_o = 2R_I$$

$$R_o = R_I$$

Max power at R_I: $R_I = R_o$ Note: R_o is usually constant or
Max voltage drop R_I: $R_I \gg R_o$ fixed by the apparatus.
Max current at R_I: $R_I \ll R_o$

Review questions

1. Using Kirchhoff's laws, find the current in the 4 Ω resistor in the network below:

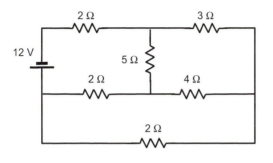

(Ans: 0.122 A)

2. Find the current through the load resistor R_L when it takes the following values:

    (a) 650 Ω
    (b) 1150 Ω
    (c) 1650 Ω
    (d) 3650 Ω

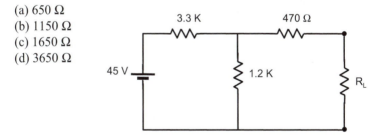

Hint: use Thevenin's theorem.

(Ans: 6, 4.8, 4, 2.4 mA)

3. A Wheatstone's bridge is used to measure temperature with the aid of a temperature sensitive resistor (a thermistor). If the meter (G) across the bridge has a resistance of 1200 Ω, and the resistance of the thermistor changes from 1500 Ω to 1600 Ω for a change in temperature of 60 °C to 61 °C, determine the change in current through the meter.

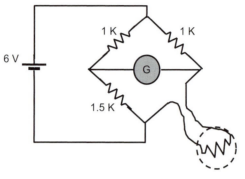

Temperature
sensitive resistor
(thermistor)

(Ans: 0.038 mA)

3. AC Circuits

Summary

$$v = V_o \sin(\omega t)$$ Instantaneous voltage

$$V_{rms} = \frac{V_o}{\sqrt{2}} = 0.707 V_o$$ Rms voltage

$$I_{rms} = \frac{I_o}{\sqrt{2}} = 0.707 I_o$$ Rms current

$$X_C = \frac{1}{\omega C}$$ Capacitive reactance

$$X_L = \omega L$$ Inductive reactance

$$P_R = V_{rms} I_{rms}$$ Reactive power

$$P_{av} = V_{rms} I_{rms} \cos \phi$$ Average (active) power

$$S = V_{rms} I_{rms}$$ Apparent power

$$|Z| = \sqrt{R^2 + (X_L - X_C)^2}$$

$$\tan \phi = \left[\frac{X_L - X_C}{R} \right]$$ Impedance

$$Z = R + j(X_L - X_C)$$
$$= R + j\left(\omega L - \frac{1}{\omega C} \right)$$

$$\frac{V_{out}}{V_{in}} = \frac{1}{\sqrt{1 + R^2 \omega^2 C^2}}$$ Low pass filter

$$\frac{V_{out}}{V_{in}} = \frac{R \omega C}{\sqrt{R^2 \omega^2 C^2 + 1}}$$ High pass filter

$$R \omega C = 1$$ 3dB point

3.1 AC Voltage

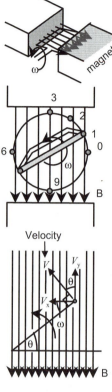

Consider a constant angular speed (ω):
- at (0) motion of conductor is parallel to B, hence induced voltage = 0;
- at (1), conductor has begun to cut magnetic field lines B, hence some voltage is induced;
- at (2), conductor cuts magnetic field lines at a greater rate than (1) and thus a greater voltage is induced;
- at (3), conductor cuts magnetic field lines at maximum rate, thus maximum voltage is induced;
- from (3) to (6), the rate of cutting becomes less.
- at (6), conductor moves parallel to B and v = 0;
- from (6) to (9), conductor begins to cut field lines again but in the opposite direction, hence, induced voltage is reversed in polarity.

The **induced voltage** is directly proportional to the rate at which the conductor cuts across the magnetic field lines. Thus, the induced voltage is proportional to the velocity of the conductor in the x direction ($V_x = V\sin\theta$). The velocity component V_y is parallel to the field lines and thus does not contribute to the rate of "cutting".

RH rule: • fingers: direction of field
         • thumb: direction of Vx
         • palm: force on *positive* charge carriers
         Thus, current is coming out from the page.

$$v_{induced} = V_o \sin \theta$$

⎿ maximum (peak) voltage V_o induced at $\theta = \pi/2$

V_o depends on:
- total number of flux lines through which the conductor passes
- angular velocity of loop
- no. turns of conductor in loop

since $\quad \omega = \dfrac{\theta}{t}$ radians

then $\quad \boxed{v = V_o \sin(\omega t)}$

Instantaneous voltage ⎯⎯ ⎿ Peak voltage

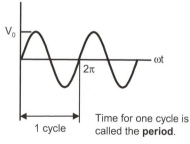

Time for one cycle is called the **period**.

1 cycle

3.2 Resistance

The **instantaneous voltage** across the resistor is:

$$v_r = V_o \sin \omega t$$

⮑ Maximum value of V

The **instantaneous current** in the resistor is:

$$i = \frac{V}{R}$$

$$= \frac{V_o}{R} \sin \omega t$$

The maximum current in the resistor is when $\sin \omega t = 1$ thus:

$$I_o = \frac{V_o}{R}$$

$$\therefore i = I_o \sin \omega t$$

Instantaneous power:

$$p = vi$$

$$= i^2 R$$

$$= (I_o \sin \omega t)^2 R$$

$$= I_o^2 R \sin^2 \omega t$$

$$P_o = I_o^2 R$$

$$p = P_o \sin^2 \omega t$$

• power is a function of $\sin^2 \omega t$
• power is sinusoidal in nature with a frequency of twice the instantaneous current and voltage and is always positive indicating power continuously supplied to the resistor.

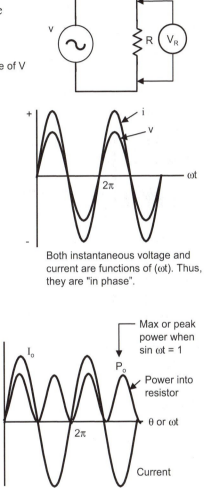

Both instantaneous voltage and current are functions of (ωt). Thus, they are "in phase".

Max or peak power when $\sin \omega t = 1$

Power into resistor

Current

Resistance is the opposition to alternating current due to the motion of charge carriers within the resistor. The opposition tendered depends upon the magnitude of voltage across the resistor.

3.3 rms voltage and current

The area under the power vs time function is energy. Thus, it is possible to calculate an **average power** level which, over one cycle, is associated with the amount of energy carried in one cycle of alternating power.

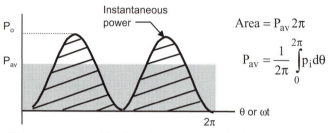

Area $= P_{av} 2\pi$

$$P_{av} = \frac{1}{2\pi} \int_0^{2\pi} p_i d\theta$$

This energy would be that given by an equivalent DC, or steady-state, voltage and current over a certain time period compared to that from an alternating current and voltage for the same time period.

Average power:

$$P_{av} = \frac{\int_0^{2\pi} p_i d\theta}{2\pi}$$

$$= \frac{1}{2\pi} \int_0^{2\pi} i^2 R d\theta$$

$$= \frac{R}{2\pi} \int_0^{2\pi} i^2 d\theta$$

$$= \frac{R}{2\pi} \int_0^{2\pi} I_o^2 \sin^2 \theta d\theta$$

$$= \frac{I_o^2 R}{2\pi} \int_0^{2\pi} \sin^2 \theta d\theta$$

$$= \frac{I_o^2 R}{2}$$

this integral evaluates to just π

Now: $P_{av} = \frac{I_o^2}{2} R$

What equivalent **steady-state** current would give the same average power as an alternating current?

Let: $I_{rms} = \dfrac{I_o}{\sqrt{2}}$ This result is only for sinusoidal signals.

Thus: $P_{av} = I_{rms}^2 R$ For resistor circuit only. See later for LCR series circuit.

or
$P_{av} = I_{rms} V_{rms}$

$= \dfrac{I_o}{\sqrt{2}}$

$= 0.707 I_0$

$= \dfrac{V_0}{\sqrt{2}}$

$= 0.707 V_0$

I_{rms} and V_{rms} are equivalent steady-state values which give the same power dissipation as the application of an alternating current with peak values V_p and I_p

In general:

$$V_{rms} = \sqrt{\frac{1}{T} \int_0^T v^2(t) dt} \qquad I_{rms} = \sqrt{\frac{1}{T} \int_0^T i^2(t) dt}$$

In AC circuits, V and I without subscripts indicate rms values unless stated otherwise.

3.4 Capacitive reactance

The AC source supplies an alternating voltage v. This
voltage appears across the capacitor.

In general, $C = \dfrac{Q}{V}$ (q,v are instantaneous
values and thus

thus $q = Cv$ functions of t)

$\dfrac{dq}{dt} = C\dfrac{dv}{dt}$ differentiating
w.r.t. time

$i = C\dfrac{dv}{dt}$ C is a "constant"

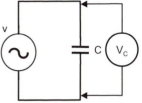

Instantaneous current is proportional to the rate of change of voltage.

The instantaneous current is a
maximum I_o when the rate of
change of voltage is a
maximum. Also, the
maximum voltage V_o
only appears across the
capacitor *after* it has
become charged
whereupon the current
I drops to zero. Thus,
maximums and minimums in
the instantaneous current lead
the maximums and minimums
in the instantaneous voltage by
$\pi/2$.

Maximums in
current in
capacitor
precede
maximums in
voltage across it.

Now, $v = V_o \sin(\omega t)$

$i = C\dfrac{d}{dt}V_o \sin(\omega t)$ since $i = C\dfrac{dv}{dt}$

thus $i = \omega C V_o \cos(\omega t)$

$= [\omega C V_o]\sin\left(\omega t + \dfrac{\pi}{2}\right)$

$I_o = \omega C V_o$ @ $i = I_o$

$i = I_o \sin\left(\omega t + \dfrac{\pi}{2}\right)$

$\dfrac{1}{\omega C} = \dfrac{V_o}{I_o}$ Can be peak or rms
but not instantaneous

$= X_C$

Capacitive
reactance (Ω)

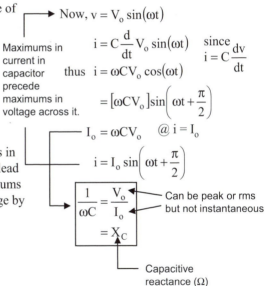

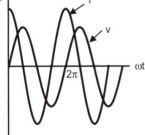

Capacitive reactance is the opposition to
alternating current by capacitance. The
opposition tendered depends upon the rate of
change of <u>voltage</u> across the capacitor.

What is capacitive reactance? How can a capacitor offer a **resistance** to alternating current?

Consider a capacitor connected to a DC supply so that the polarity of the applied voltage can be reversed by a switch.

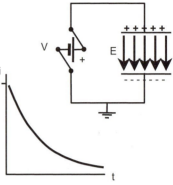

When the switch is first closed, it takes time for the charge Q to accumulate on each plate. Charge accumulation proceeds until the voltage across the capacitor is equal to the voltage of the source. During this time, current flows in the circuit.

When the polarity is reversed, the capacitor initially discharges and then charges to the opposite polarity. Current flows in the opposite direction while reverse charging takes place until the voltage across the capacitor becomes equal to the supply voltage whereupon current flow ceases.

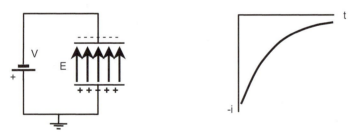

Now, if the switch were to be operated very quickly, then, upon charging, the current would not have time to drop to zero before the polarity of the supply voltage was reversed. Similarly, on reverse charging, the reverse current would not have time to reach zero before the polarity of the source was reversed. Thus, the current would only proceed a short distance along the curves as shown and a continuous alternating current would result. The faster the switch over of polarity, the greater the average or rms AC current. Thus, the "resistance" to AC current is greater at lower frequencies and lower at high frequencies.

3.5 Inductive reactance

Let the inductor have no resistance. Thus, any voltage that appears across the terminals of the inductor must be due to the self-induced voltage in the coil by a changing current through it (self-inductance).

At any instant, $v = v_L$ by Kirchhoff $\longrightarrow$ Note: if we had included the – sign for $L di/dt$, then we would be treating v_L as a voltage source of opposite polarity to v and it would appear on the left hand side of the equation.

$$v_L = L\frac{di}{dt}$$

$$v = V_p \sin(\omega t)$$ $\longleftarrow$

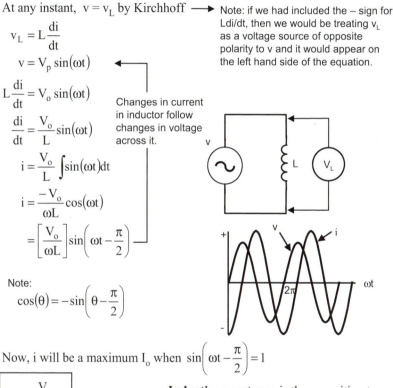

$$L\frac{di}{dt} = V_o \sin(\omega t)$$

Changes in current in inductor follow changes in voltage across it.

$$\frac{di}{dt} = \frac{V_o}{L}\sin(\omega t)$$

$$i = \frac{V_o}{L}\int \sin(\omega t)dt$$

$$i = \frac{-V_o}{\omega L}\cos(\omega t)$$

$$= \left[\frac{V_o}{\omega L}\right]\sin\left(\omega t - \frac{\pi}{2}\right)$$

Note:
$$\cos(\theta) = -\sin\left(\theta - \frac{\pi}{2}\right)$$

Now, i will be a maximum I_o when $\sin\left(\omega t - \frac{\pi}{2}\right) = 1$

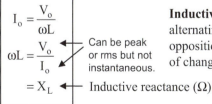

$$I_o = \frac{V_o}{\omega L}$$

$$\omega L = \frac{V_o}{I_o}$$ $\longleftarrow$ Can be peak or rms but not instantaneous.

$$= X_L$$ $\longleftarrow$ Inductive reactance (Ω)

Inductive reactance is the opposition to alternating current by inductance. The opposition tendered depends upon the rate of change of <u>current</u> through the inductor.

For high frequencies, the magnitude of the induced back emf is large and this restricts the maximum current that can flow before the polarity of the supply voltage changes over. Thus, the reactance increases with increasing frequency.

3.6 LCR series circuit

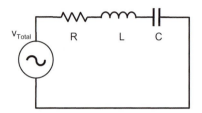

A varying voltage v_{Total} from the source will cause a varying instantaneous current i to flow in the circuit. Because it is a series circuit, the current must be the same in each part of the circuit at any particular time t.

- For the resistance, changes in v_R are in phase with those of i
- For the inductor, changes in v_L are ahead of those of i by $\pi/2$
- For the capacitor, changes in v_C follow those of i by $\pi/2$

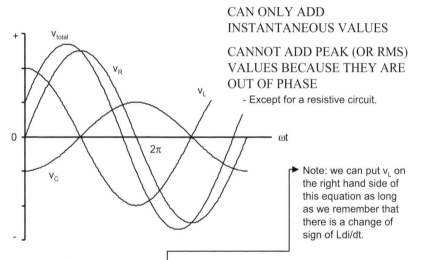

CAN ONLY ADD INSTANTANEOUS VALUES

CANNOT ADD PEAK (OR RMS) VALUES BECAUSE THEY ARE OUT OF PHASE
- Except for a resistive circuit.

Note: we can put v_L on the right hand side of this equation as long as we remember that there is a change of sign of Ldi/dt.

From Kirchhoff, $v_{Total} = v_R + v_L + v_C$ at any instant. Note that each of these voltages do not reach their peak values when v_{Total} reaches a maximum, thus $|V_{oTotal}| <> |V_{oR}| + |V_{oL}| + |V_{oC}|$. Also, since the rms value of any voltage $= 0.707 \, V_o$, then $|Vrms_{total}| <> |Vrms_R| + |Vrms_L| + |Vrms_C|$

Algebraic addition generally only applies to instantaneous quantities (can be applied to other quantities, e.g. peak or rms, if current and voltage are in phase - such as resistor circuit only).

Note also that the resultant of the addition of sine waves of same frequency is a sine wave of same frequency.

3.7 LCR circuit – peak and rms voltage

In LCR series circuits, how then may the total or <u>resultant</u> peak (or rms) values of voltage and current be determined from the individual peak (or rms) voltages? A VECTOR approach is needed (can use **complex numbers**).

Consider the axes below which indicates either peak (or rms) voltages V_R, V_C and V_L.

┌ peak or rms

V_L

V_{Total}

φ

V_R

Current is common
point of reference in
series circuit.

V_C

In a *series* circuit, the current is the same in each component. The instantaneous voltage across the resistor is always in phase with the instantaneous current. For the inductor and the capacitor:

• peak or rms values of V_L always precede V_R by $\pi/2$ and V_L is drawn upwards on the vertical axis;

• peak or rms values of V_C always follow V_R by $\pi/2$ and V_C is downwards on the vertical axis.

The resultant peak or rms voltage V_T is the <u>vector</u> sum of $V_R + V_L + V_C$

The angle φ is the phase angle of the resultant peak (or rms) voltage w.r.t. the peak (or rms) common current and is found from:

$$\tan \phi = \frac{V_L - V_C}{V_R}$$

For an AC series circuit:

• same current flows in all components;
• vector sum of rms or peak voltages must equal the applied rms or peak voltage;
• algebraic sum of instantaneous voltages equals the applied instantaneous voltage.

In **complex number** form:

$$V_T = V_R + j(V_L - V_C)$$

Complex numbers are a convenient mathematical way to keep track of directions or "phases" of quantities.

3.8 Impedance

The total opposition to current in an AC circuit is called **impedance**. For a series circuit, the impedance is the vector sum of the resistances and the reactances within the circuit

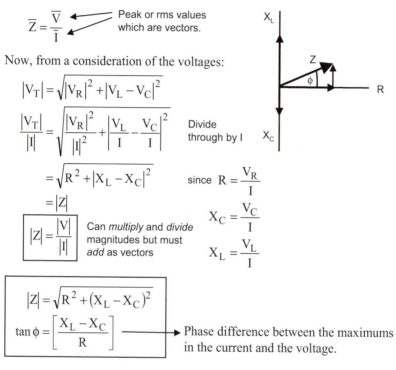

$$\overline{Z} = \frac{\overline{V}}{\overline{I}}$$ ← Peak or rms values which are vectors.

Now, from a consideration of the voltages:

$$|V_T| = \sqrt{|V_R|^2 + |V_L - V_C|^2}$$

$$\frac{|V_T|}{|I|} = \sqrt{\frac{|V_R|^2}{|I|^2} + \left|\frac{V_L}{I} - \frac{V_C}{I}\right|^2}$$ Divide through by I

$$= \sqrt{R^2 + |X_L - X_C|^2}$$ since $R = \dfrac{V_R}{I}$

$$= |Z|$$

$$\boxed{|Z| = \frac{|V|}{|I|}}$$ Can *multiply* and *divide* magnitudes but must *add* as vectors

$$X_C = \frac{V_C}{I}$$

$$X_L = \frac{V_L}{I}$$

$$\boxed{\begin{array}{l} |Z| = \sqrt{R^2 + (X_L - X_C)^2} \\[2mm] \tan\phi = \left[\dfrac{X_L - X_C}{R}\right] \end{array}}$$ ⟶ Phase difference between the maximums in the current and the voltage.

For an RC series circuit: For an RL series circuit:

$$\tan\phi = \left[\frac{-X_C}{R}\right]$$ $$\tan\phi = \left[\frac{X_L}{R}\right]$$

$$= \frac{-1}{R\omega C}$$ $$= \frac{\omega L}{R}$$

If $X_C > X_L$, circuit is **capacitively reactive**.
If $X_C < X_L$, circuit is **inductively reactive**.
If $X_C = X_L$, circuit is **resonant**.

3. 9 Low pass filter

In a low pass filter, low frequencies let through: high frequencies attenuated.

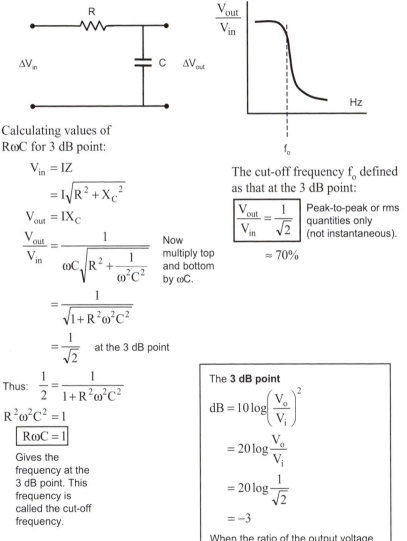

Calculating values of $R\omega C$ for 3 dB point:

$$V_{in} = IZ$$

$$= I\sqrt{R^2 + X_C^2}$$

$$V_{out} = IX_C$$

$$\frac{V_{out}}{V_{in}} = \frac{1}{\omega C\sqrt{R^2 + \dfrac{1}{\omega^2 C^2}}}$$

Now multiply top and bottom by ωC.

$$= \frac{1}{\sqrt{1 + R^2 \omega^2 C^2}}$$

$$= \frac{1}{\sqrt{2}} \quad \text{at the 3 dB point}$$

Thus: $\dfrac{1}{2} = \dfrac{1}{1 + R^2 \omega^2 C^2}$

$$R^2 \omega^2 C^2 = 1$$

$$\boxed{R\omega C = 1}$$

Gives the frequency at the 3 dB point. This frequency is called the cut-off frequency.

The cut-off frequency f_o defined as that at the 3 dB point:

$$\boxed{\frac{V_{out}}{V_{in}} = \frac{1}{\sqrt{2}}}$$

Peak-to-peak or rms quantities only (not instantaneous).

$$\approx 70\%$$

The **3 dB point**

$$dB = 10 \log \left(\frac{V_o}{V_i} \right)^2$$

$$= 20 \log \frac{V_o}{V_i}$$

$$= 20 \log \frac{1}{\sqrt{2}}$$

$$= -3$$

When the ratio of the output voltage to the input voltage is $1/\sqrt{2}$, then this corresponds to a drop of 3 dB.

3.10 High pass filter

In a high pass filter, high frequencies are let through: low frequencies are attenuated.

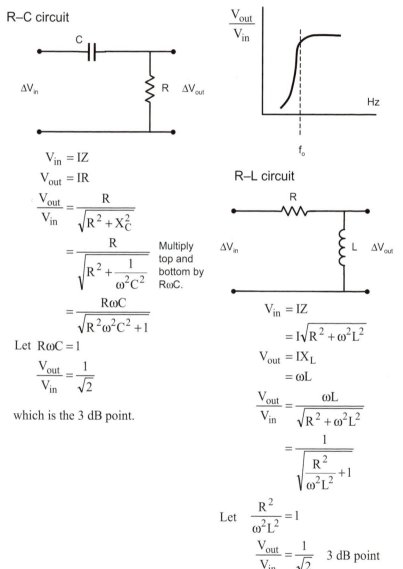

R–C circuit

$$V_{in} = IZ$$

$$V_{out} = IR$$

$$\frac{V_{out}}{V_{in}} = \frac{R}{\sqrt{R^2 + X_C^2}}$$

$$= \frac{R}{\sqrt{R^2 + \dfrac{1}{\omega^2 C^2}}}$$

Multiply top and bottom by $R\omega C$.

$$= \frac{R\omega C}{\sqrt{R^2 \omega^2 C^2 + 1}}$$

Let $R\omega C = 1$

$$\frac{V_{out}}{V_{in}} = \frac{1}{\sqrt{2}}$$

which is the 3 dB point.

R–L circuit

$$V_{in} = IZ$$

$$= I\sqrt{R^2 + \omega^2 L^2}$$

$$V_{out} = IX_L$$

$$= \omega L$$

$$\frac{V_{out}}{V_{in}} = \frac{\omega L}{\sqrt{R^2 + \omega^2 L^2}}$$

$$= \frac{1}{\sqrt{\dfrac{R^2}{\omega^2 L^2} + 1}}$$

Let $\dfrac{R^2}{\omega^2 L^2} = 1$

$$\frac{V_{out}}{V_{in}} = \frac{1}{\sqrt{2}} \quad \text{3 dB point}$$

3.11 Complex impedance (series)

For a series circuit, the **impedance** is the vector sum of the resistances and the reactances within the circuit.

(V, I can be either peak or rms. Note, all quantities are vectors.)

$$Z = \frac{V}{I} \quad \text{by definition}$$

including Z

$$V = V_R + j(V_L - V_C)$$

$$\frac{V}{I} = \frac{V_R}{I} + j\frac{(V_L - V_C)}{I}$$

but

$$R = \frac{V_R}{I}$$

$$X_C = \frac{V_C}{I}$$

$$X_L = \frac{V_L}{I}$$

thus

$$Z = R + j(X_L - X_C)$$

$$= R + j\left(\omega L - \frac{1}{\omega C}\right)$$

Note:

$$|Z| = \frac{|V|}{|I|}$$

Magnitudes of the peaks or rms voltages. For series or parallel circuits, cannot simply add peak or rms values.

If $X_C > X_L$, circuit is **capacitively reactive**.
If $X_C < X_L$, circuit is **inductively reactive**.
If $X_C = X_L$ - then **resonant**.

$X = X_L - X_C$

Modulus of Z

$$|Z| = \sqrt{R^2 + (X_L - X_C)^2}$$

$$\tan\phi = \left[\frac{X_L - X_C}{R}\right]$$

Phase difference between the current and the voltage.

e.g. for an RC series circuit,

$$\tan\phi = \left[\frac{-X_C}{R}\right]$$

$$= \frac{-1}{R\omega C}$$

for an RL series circuit,

$$\tan\phi = \left[\frac{X_L}{R}\right]$$

$$= \frac{\omega L}{R}$$

3.12 Resonance (series)

Consider a series LCR circuit where:

$$X_C = \frac{1}{\omega C}$$

$$X_L = \omega L$$

the resonant frequency ω_R

and at some frequency, $X_C = X_L$.

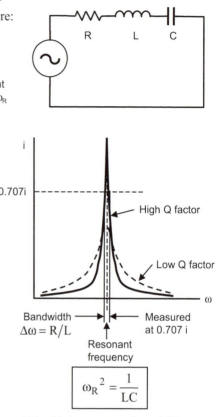

$$\omega_R L = \frac{1}{\omega_R C}$$

$$\omega_R{}^2 = \frac{1}{LC}$$

$$\omega_R = \frac{1}{\sqrt{LC}} \quad \text{condition for resonance.}$$

At the **resonant frequency**, with $X_L = X_C$, the **impedance** Z will be a minimum.

$$Z = R + j(X_L - X_C)$$

The current i will then be a maximum (and in phase with the voltage v).

The **Q factor** is an indication of the sharpness of the peak. High Q indicates sharp peak, low Q broad peak.

0.707i

High Q factor

Low Q factor

Bandwidth
$\Delta\omega = R/L$

Measured at 0.707 i

Resonant frequency

$$\omega_R{}^2 = \frac{1}{LC}$$

Note: this is resonance for an LCR series circuit. L, C and R must be present for resonance to occur. If R = 0, then resonance peak is infinitely high.

$$Q = \frac{\omega_R L}{R} = \frac{1}{R\omega_R C}$$

$$= \frac{1}{R}\sqrt{\frac{L}{C}}$$

The Q factor describes the **selectivity** of the circuit as well as the magnification of the voltage across the inductor and the capacitor.
At resonance, $V_C = QV$, $V_L = QV$

3.13 Impedance (parallel)

A **parallel circuit** is one in which the same voltage appears across all components. For parallel circuits, the voltage is the common point of reference (rather than the current as was the case in series circuits).

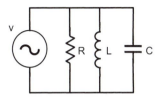

In a parallel circuit, the instantaneous current across the resistor is always in phase with the instantaneous voltage, thus, for the capacitor and the inductor:

peak or rms

- I_C always precedes I_R by $\pi/2$ thus I_C is upwards on the vertical axis;

- I_L always follows I_R by $\pi/2$ thus I_L is downwards on the vertical axis.

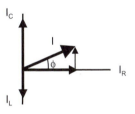

Voltage is the common point of reference for a parallel circuit:

- same voltage across all components;
- vector sum of rms or peak currents must equal the rms or peak current;
- algebraic sum of instantaneous currents equals the total instantaneous current.

In complex number form:

$$\frac{1}{Z} = \frac{1}{R} + j\left(\frac{1}{X_C} - \frac{1}{X_L}\right)$$

$Y = 1/Z$ is **admittance** (units: siemens).

Note: this formula is consistent with addition of parallel impedances Z. The "j" has been moved to numerator by multiplying through by j/j and remembering that $j^2 = -1$ (hence positions of X_L and X_C reversed).

At **resonance**, $X_C = X_L$ and total current i is a minimum since Z is a maximum.

$$\omega_R^2 = \frac{1}{LC}$$

Parallel circuits exhibit high impedance at resonance.

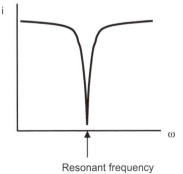

Resonant frequency

3.14 Impedances (series and parallel)

Impedances must always be added as vectors.

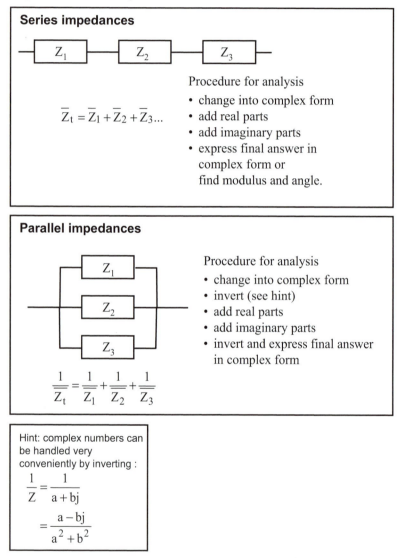

Series impedances

$$\overline{Z}_t = \overline{Z}_1 + \overline{Z}_2 + \overline{Z}_3 ...$$

Procedure for analysis
* change into complex form
* add real parts
* add imaginary parts
* express final answer in complex form or find modulus and angle.

Parallel impedances

$$\frac{1}{\overline{Z}_t} = \frac{1}{\overline{Z}_1} + \frac{1}{\overline{Z}_2} + \frac{1}{\overline{Z}_3}$$

Procedure for analysis
* change into complex form
* invert (see hint)
* add real parts
* add imaginary parts
* invert and express final answer in complex form

Hint: complex numbers can be handled very conveniently by inverting :

$$\frac{1}{Z} = \frac{1}{a + bj}$$

$$= \frac{a - bj}{a^2 + b^2}$$

3.15 Impedances (example)

Determine the **total impedance** of this circuit at a frequency $\omega = 3$ kHz
(18850 rad s^{-1})

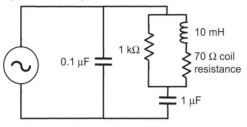

1. Express each component in complex form

0.1 µF $Z = 0 + j\left(-18850\left(0.1 \times 10^{-6}\right)\right)^{-1}$
         $= 0 + -530j$

1 µF $Z = 0 + j\left(-18850\left(1 \times 10^{-6}\right)\right)^{-1}$
         $= 0 + -53j$

10 mH $Z = 70 + j(18850)\left(10 \times 10^{-3}\right)$
         $= 70 + 188.5j$

1 kΩ $Z = 1000 + j(0)$

2. Combine inductor and
   resistor

$$\frac{1}{Z} = \frac{1}{1000} + \frac{1}{70 + 188.5j}$$

$$= \frac{1}{1000} + \frac{70 - 188.5j}{(70 + 188.5j)(70 - 188.5j)}$$

$$= \frac{1}{1000} + \frac{70 - 188.5j}{4900 + 35532}$$

$$= 2.73 \times 10^{-3} - 4.7 \times 10^{-3} j$$

$$Z = \frac{2.73 \times 10^{-3} + 4.7 \times 10^{-3} j}{\left(2.73 \times 10^{-3}\right)^2 + \left(4.7 \times 10^{-3}\right)^2}$$

$$= 92.4 + 159j$$

3. Combine with
   capacitor in series
   $$Z = 92.4 + (159 - 53)j$$
   $$= 92.4 + 106j$$

4. Combine with capacitor
   in parallel

$$\frac{1}{Z} = \frac{1}{-530j} + \frac{1}{92.4 + 106j}$$

$$= \frac{530j}{280900} + \frac{92.4 - 106j}{19773}$$

$$= 4.67 \times 10^{-3} - 3.46 \times 10^{-3} j$$

$$Z = \frac{4.67 \times 10^{-3} + 3.46 \times 10^{-3} j}{3.38 \times 10^{-5}}$$

$$= 138 + 102.4j\,\Omega$$

$$\boxed{Z = 171.8\Omega,\ 36.6°}$$

3.16 AC circuits

Series AC circuit

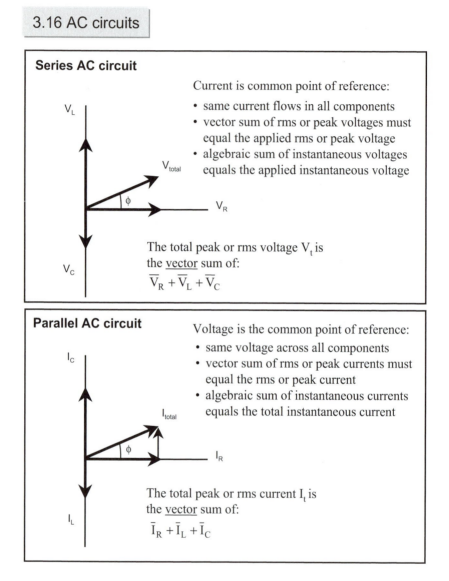

Current is common point of reference:
- same current flows in all components
- vector sum of rms or peak voltages must equal the applied rms or peak voltage
- algebraic sum of instantaneous voltages equals the applied instantaneous voltage

The total peak or rms voltage V_t is the <u>vector</u> sum of:

$$\overline{V}_R + \overline{V}_L + \overline{V}_C$$

Parallel AC circuit

Voltage is the common point of reference:
- same voltage across all components
- vector sum of rms or peak currents must equal the rms or peak current
- algebraic sum of instantaneous currents equals the total instantaneous current

The total peak or rms current I_t is the <u>vector</u> sum of:

$$\overline{I}_R + \overline{I}_L + \overline{I}_C$$

3.17 AC circuits (example)

Calculate the **impedance** of the network shown at $\omega = 1000$ rad s^{-1} and also the current in the 20 μF capacitor.

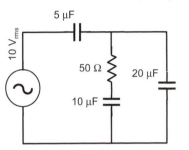

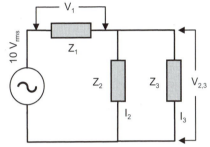

1. By methods of previous example, determine total impedance:

$$Z_1 = 0 - 200\text{j}$$
$$Z_2 = 50 - 100\text{j}$$
$$Z_3 = 0 - 50\text{j}$$
$$Z_2 \parallel Z_3 = 5 - 35\text{j}$$
$$Z_T = 5 - 235\text{j}$$
$$V = IZ$$
$$I = \frac{10}{5 - 235\text{j}}$$
$$= 9 \times 10^{-4} + 0.0425\text{j}$$

Note: all quantities are vectors

2. Then apply Kirchhoff's laws:

$$V_1 = \left(9 \times 10^{-4} + 0.0425\text{j}\right)\left(0 - 200\text{j}\right)$$
$$= 8.5 - 0.181\text{j}$$
$$V_{2,3} = \left(9 \times 10^{-4} + 0.0425\text{j}\right)\left(5 - 35\text{j}\right)$$
$$= 1.488 + 0.181\text{j}$$
$$|V_3| = 1.5$$
$$V_1 + V_{2,3} = 10 + 0\text{j}$$
$$V_3 = I_3 Z_3$$
$$I_3 = \frac{1.488 + 0.181\text{j}}{0 - 50\text{j}}$$
$$= -3.62 \times 10^{-3} + 0.02976\text{j}$$
$$|I_3| = 0.03\text{A}$$

or

$$1.5 = I_{rms}|Z_3|$$
$$I_{rms} = 0.03 \text{ A}$$

Note: can do multiplication with magnitudes but not additions.

3.18 Filters: complex form

Low pass filter

$$V_{in} = IZ$$
$$= I(R - X_C j)$$
$$= I\left(R - \frac{1}{\omega C}j\right)$$
$$V_{out} = I(-X_C j)$$
$$= I\left(-\frac{1}{\omega C}j\right)$$

$$\frac{V_{out}}{V_{in}} = \frac{-\dfrac{1}{\omega C}j}{R - \dfrac{1}{\omega C}j}$$

⇩

$$= \frac{1 - R\omega Cj}{1 + R^2\omega^2 C^2}$$

$$= \frac{1}{1 + R\omega Cj}$$

$$\left|\frac{V_{out}}{V_{in}}\right| = \frac{1}{\sqrt{1^2 + R^2\omega^2 C^2}}$$

Let $R\omega C = 1$

$$\frac{V_{out}}{V_{in}} = \frac{1}{\sqrt{2}} \quad \text{which is the } 3\text{ dB point}$$

High pass filter

Peak-to-peak or rms
quantities only.

$$V_{in} = I(R + -X_C j)$$
$$V_{out} = IR$$

$$\frac{V_{out}}{V_{in}} = \frac{R}{R - X_c j}$$

$$\frac{R}{R - \dfrac{1}{\omega C}j} = \frac{R}{\sqrt{R^2 + \dfrac{1}{\omega^2 C^2}}}$$

$$\left|\frac{V_{out}}{V_{in}}\right| = \frac{R\omega C}{\sqrt{R^2\omega^2 C^2 + 1}}$$

Let $R\omega C = 1$

$$\frac{V_{out}}{V_{in}} = \frac{1}{\sqrt{2}} \quad \text{which is the } 3\text{ dB point}$$

For a fixed R and C, the frequency at the 3 dB point is called the **cut-off frequency** f_o at which point:

$$V_{out} = \frac{V_{in}}{\sqrt{2}}$$

Peak-to-peak or rms
quantities only
(not instantaneous)

3.19 Signal generator and oscilloscope

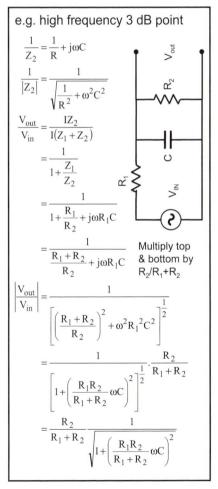

The input and output impedances of signal generating and measuring devices has a marked effect on the attenuation of voltage signals that may be produced or measured. For example, an oscillator with an output impedance of 600 Ω may be connected to the input of an oscilloscope with an input resistance of 2 kΩ and a series capacitance of 10 μF (on the AC setting of the CRO). The capacitance of the signal leads shunts the input of the oscilloscope and may be around 20 pF. At low frequencies, the shunt capacitance may be ignored. At high frequencies, the series capacitance may be ignored. In the mid-frequency range, we may assume that both capacitances have a negligible effect on the signal.

By determining the 3 dB points for the equivalent high and low frequency circuits, two frequencies f_{lo} and f_{hi} can be found at which the input voltage ΔV_{in} drops by 3 dB relative to its mid-frequency value (assuming that ΔV_s is constant at all frequencies).

e.g. high frequency 3 dB point

$$\frac{1}{Z_2} = \frac{1}{R} + j\omega C$$

$$\frac{1}{|Z_2|} = \frac{1}{\sqrt{\frac{1}{R^2} + \omega^2 C^2}}$$

$$\frac{V_{out}}{V_{in}} = \frac{I Z_2}{I(Z_1 + Z_2)}$$

$$= \frac{1}{1 + \frac{Z_1}{Z_2}}$$

$$= \frac{1}{1 + \frac{R_1}{R_2} + j\omega R_1 C}$$

$$= \frac{1}{\frac{R_1 + R_2}{R_2} + j\omega R_1 C}$$

Multiply top & bottom by $R_2/R_1 + R_2$

$$\left|\frac{V_{out}}{V_{in}}\right| = \frac{1}{\left[\left(\frac{R_1 + R_2}{R_2}\right)^2 + \omega^2 R_1^2 C^2\right]^{\frac{1}{2}}}$$

$$= \frac{1}{\left[1 + \left(\frac{R_1 R_2}{R_1 + R_2}\omega C\right)^2\right]^{\frac{1}{2}}} \cdot \frac{R_2}{R_1 + R_2}$$

$$= \frac{R_2}{R_1 + R_2} \frac{1}{\sqrt{1 + \left(\frac{R_1 R_2}{R_1 + R_2}\omega C\right)^2}}$$

3.20 AC bridge

For the galvanometer to read zero, the voltage across its terminals must be zero. Thus, the voltage across Z_1 and Z_3 must be equal in magnitude and phase.

$$V_1 = V_3$$
$$I_1 Z_1 = I_3 Z_3$$

At balance condition, no current flows through galvanometer

thus $V = I_1(Z_1 + Z_2)$
$$= I_3(Z_3 + Z_4)$$

$$I_1 = \frac{V}{Z_1 + Z_2}; \quad I_3 = \frac{V}{Z_3 + Z_4}$$

$$I_1 Z_1 = \frac{Z_1 V}{Z_1 + Z_2}; \quad I_3 Z_3 = \frac{Z_3 V}{Z_3 + Z_4}$$

$$\frac{Z_3 V}{Z_3 + Z_4} = \frac{Z_1 V}{Z_1 + Z_2}$$

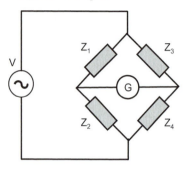

⬇

$$\boxed{Z_1 Z_4 = Z_2 Z_3}$$ **General bridge equation** at balance condition

Example

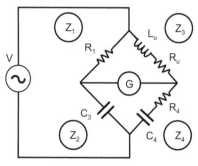

$$Z_1 = R_1$$
$$Z_2 = 0 - \frac{1}{\omega C_3 j}$$
$$Z_3 = R_u + \omega L_u$$
$$Z_4 = R_4 - \frac{1}{\omega C_4}$$

$$Z_1 Z_4 = R_1 \left(R_4 - \frac{1}{\omega C_4} j \right)$$
$$= R_1 R_4 - \frac{R_1}{\omega C_4} j$$

$$Z_2 Z_3 = \left(0 - \frac{1}{\omega C_3} j \right)(R_u + \omega L_u j)$$
$$= \frac{L_u}{C_3} + \frac{-R_u}{\omega C_3} j$$

$$R_1 R_4 - \frac{R_1}{\omega C_4} j = \frac{L_u}{C_3} + \frac{-R_u}{\omega C_3} j$$

$$\boxed{R_1 R_4 = \frac{L_u}{C_3}}$$

$$\boxed{\frac{R_1}{C_4} = \frac{R_u}{C_3}}$$

Review questions

1. Determine the rms voltages of the waveforms shown below from first principles. In both cases, assume that the circuits are resistive (voltage and current are in phase).

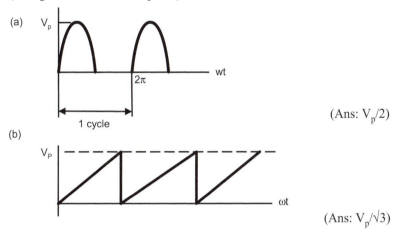

(a)

(Ans: $V_p/2$)

(b)

(Ans: $V_p/\sqrt{3}$)

2. The average (i.e. rms) power of a soldering iron is 55 W and the iron has a resistance of 250 Ω. Calculate the (a) rms current, (b) rms voltage and (c) peak voltage. If the power supply in is 240 V rms, determine the peak voltage V_p.

(Ans: 469 mA, 117.25 V, 166 V, 340 V)

3. An electric toaster draws 3 A rms from a 240 V, 50 Hz source. Calculate the average power and the peak value of the instantaneous power to the toaster.

(Ans: 720 W, 1440 W)

4. What is the capacitive reactance of a 47 pF capacitor when the frequency is (a) 5 MHz, (b) 1 kHz?

(Ans: 676.9 Ω, 3.39 MΩ)

5. What value of capacitor is needed to limit the rms current through it to 3 mA when it is connected across a 50 V, 500 Hz AC source?

(Ans: 19 nF)

6. The inductance of an ignition coil in a motor vehicle is 0.005 H. The resistance of its windings is 1.5 Ω. Determine the impedance of the coil when the current in the circuit turns on and off 1000 times per second.

(Ans: 1.59 Ω)

7. Determine the impedance of a 200 pF capacitor connected in series with a 1.2 kΩ resistor when the frequency is 5 MHz.

(Ans: 1210 Ω)

8. Find the magnitude and phase with respect to V of each of the following fpr the circuit shown below:
(a) current;
(b) voltage across R;
(c) voltage across L;
(d) average power supplied by source.

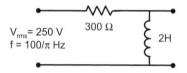

V_{rms}= 250 V
f = 100/π Hz
300 Ω
2H

(Ans: 0.5 A, −53°, 150 V, −53°, 200 V,+37°, 75 W)

9. Three impedances Z_1, Z_2 and Z_3 are connected in series. Calculate the total impedance.

$$Z_1 = 240\,\Omega, +10°$$

$$Z_2 = 40\,\Omega, +40°$$

$$Z_3 = 1000\,\Omega, -10°\quad \text{(Ans: 1256 } \Omega, -4.85°)$$

10. Determine the value of capacitor which must be connected in series with a 600 Ω resistor to limit its power dissipation to 5W when connected to a 240 V, 50 Hz source.

(Ans: 276 nF)

11. In the circuit shown, determine the frequency corresponding to the 3 dB point.

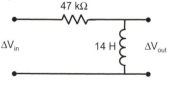

47 kΩ

ΔV_{in}
14 H
ΔV_{out}

(Ans: 534 Hz)

4. Diode

Summary

$$I = I_o \left(e^{eV/kT} - 1 \right)$$ Diode equation

$$r = \frac{25}{I}$$ Dynamic resistance
(I in mA)

4.1 Semiconductors

There are three classes of materials:

1. **Conductors**

   Valence electrons are weakly bound to the atomic lattice and are free to move about from atom to atom.

2. **Insulators**

   Valence electrons are tightly bound to the atomic lattice and are fixed in position.

3. **Semiconductors**

   In **semiconductors**, thermal vibration of atoms in a crystal causes electrons to break away and become free. The solid then becomes weakly "conducting":

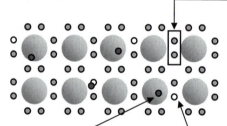

e.g. these two electrons shared equally between neighbouring atoms i.e. a covalent bond

Conductivity increases with increasing temperature.

Electrons which have broken away enter what is called the **conduction band** and have a higher energy than those left behind in the **valence band**.

The place left vacant by such an electron is called a **hole**.

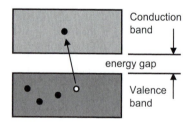

Intrinsic semiconductor

What is an energy band?

A popular model of the atom describes the structure of atoms in terms of a series of electron energy levels (1s, 2s, 2p etc). In a solid, (as distinct from a single atom) the effects of neighbouring atoms give rise to the condition that each electron energy level splits into a number of closely-spaced sub-levels. In a solid, there are many neighbouring atoms to any one atom and the total effect is for the sub-levels to become continuous and thus the energy level is more correctly called an energy band. In a metal, the valence band overlaps with the conduction band. In an insulator, there is a considerable energy gap. In a semiconductor, some electrons escape from the valence band across a small energy gap into the conduction band.

4.2 p- and n-type semiconductors

The addition of certain impurities (doping) to the silicon lattice can increase conductivity.

1. Introduction of a phosphorous atom (5 valence electrons)

P atom is still electrically neutral

The now available electron is free to wander around in the **conduction band**.

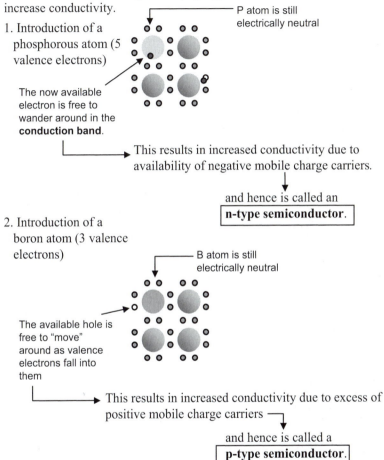

This results in increased conductivity due to availability of negative mobile charge carriers.

and hence is called an

n-type semiconductor.

2. Introduction of a boron atom (3 valence electrons)

B atom is still electrically neutral

The available hole is free to "move" around as valence electrons fall into them

This results in increased conductivity due to excess of positive mobile charge carriers

and hence is called a

p-type semiconductor.

In both types of semiconductor, the increased conductivity arises due to the deliberate increase in the number of *mobile* charge carriers - all still electrically neutral material.

- The **majority carriers** in an n-type material are electrons, the majority carriers in a p-type material are holes.
- Both types have thermally generated electrons and holes which are called **minority carriers**.

4.3 Response in an electric field

n-type (majority carriers: electrons)

1. Electrons, attracted by positive charge of battery, enter wire from conduction band.

2. Electrons from wire enter conduction band.

3. A few (intrinsic) holes drift towards negative.

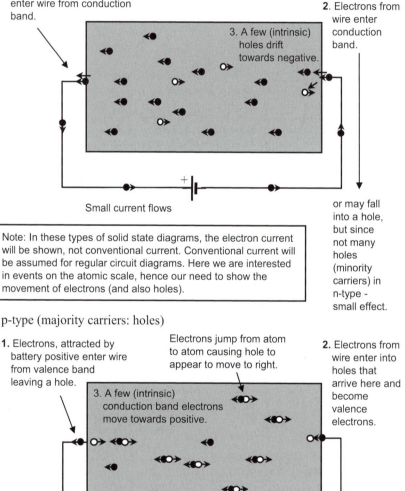

Small current flows

or may fall into a hole, but since not many holes (minority carriers) in n-type - small effect.

Note: In these types of solid state diagrams, the electron current will be shown, not conventional current. Conventional current will be assumed for regular circuit diagrams. Here we are interested in events on the atomic scale, hence our need to show the movement of electrons (and also holes).

p-type (majority carriers: holes)

1. Electrons, attracted by battery positive enter wire from valence band leaving a hole.

Electrons jump from atom to atom causing hole to appear to move to right.

2. Electrons from wire enter into holes that arrive here and become valence electrons.

3. A few (intrinsic) conduction band electrons move towards positive.

Small current flows

4.4 p-n junction

1. Near the junction, free electrons from the n side **diffuse** across the junction and fall into holes on the p side becoming valence electrons.

> Thermal energy causes free electrons to have "random motion". Excess of free electrons on the n side constitutes a "concentration gradient". These two conditions permit a net transfer of free electrons from the n side to the p side by "diffusion" … and vice versa on the p side.

2. The resulting build-up of negative charge on the p side and positive charge on the n side establishes an increasing electric field E_d across the junction leading to what is called the **barrier potential** V_b.

> V_b = 0.7 V (Si) @ 25 °C decreasing with increasing temperature.

3. Balance between diffusion process and electrostatic repulsion due to field is quickly established, no more net movement of charge carriers.

> Note, the barrier potential cannot be measured with a voltmeter due to the presence of "contact" potentials.

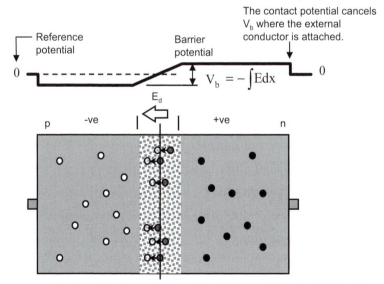

The area near the junction becomes free of majority carriers (and is therefore an insulator) and is called the **depletion region**. Any thermally generated minority carriers within the depletion region are swept across it by field E_d. Accumulation of charge is reduced somewhat and diffusion then re-establishes equilibrium and E_d resumes its former value.

4.5 Contact potential

When a metal is placed in contact with a semiconductor (or even another metal), the difference between the density of free electrons on either side of the contact, or junction, causes a **concentration gradient** which results in **diffusion** of electrons across the contact junction.

Whether or not electrons diffuse from the metal to the semiconductor, or from the semiconductor to the metal depends upon whether the semiconductor is p- or n-type and the nature of the metal.

> The "nature" of the metal and the contact is beyond the scope of our discussion. It is connected with the energy levels of the conduction electrons in each of the materials and how they compare with each other.

This movement of electrons across the contact gives rise to an electric **contact potential**. Because of contact potentials, you cannot measure the barrier potential of a p-n junction by connecting a voltmeter across it.

For the contact between the n-type semiconductor and a copper wire, electrons flow from the metal to the semiconductor. This cancels the positive potential at the end of the n-type material.

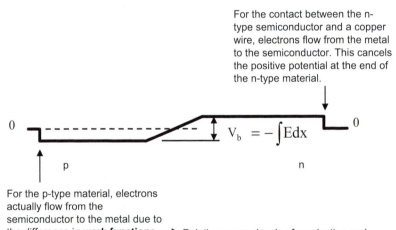

$$V_b = -\int E dx$$

For the p-type material, electrons actually flow from the semiconductor to the metal due to the difference in **work functions** of these two materials. This cancels the negative potential in the semiconductor at this contact.

Relative energy levels of conduction and valence bands in both materials. A full understanding of contact potentials requires a study of solid state physics.

4.6 Potential diagram

In this type of diagram, electrons are like marbles or round beads. Electrons are given a boost upwards by emfs or voltage sources. Electrons may only move if there is a downhill path or slope. If the slope is frictionless (i.e. an electric field), then electrons gain "kinetic" energy. If the slope has friction (i.e. resistor), then energy is converted to heat.

p-n junction

Diffusion acts like a "force" D to push electrons from the n side to the p side. As electrons accumulate on the p side, the slope becomes steeper and steeper until it is too difficult for the diffusion force (which remains constant) to transfer any more electrons from p to n.

The slope (**barrier potential**) arises due to the accumulation of charge on either side of the junction. Diffusion acts to transfer electrons up the slope. If the slope is too steep, then no electrons are transferred. Electrons cannot be transferred part-way up the slope, they either get transferred to the top or wait at the bottom.

If an **electron-hole pair** should appear on the slope (via thermal agitation) then the electron rolls downhill to the n side. This electron cancels some of the +charge on the n side thus reducing the steepness of the slope slightly. Diffusion can then transfer one more electron to the p side whereupon the steepness of the slope resumes its previous value and diffusion is not strong enough to transfer any further electrons.

4.7 Forward bias

1. Battery voltage V causes an external field E across the
    depletion layer to cancel the internal field E_d. Majority
    carriers in both materials can cross the junction because
    the applied emf overcomes the barrier potential. The
    depletion region disappears.

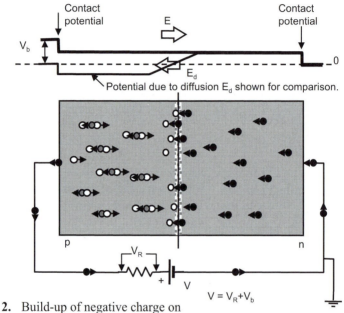

2. Build-up of negative charge on
    the p side is prevented as
    electrons drain away to battery
    positive terminal.

3. Continuous flow of electrons on the n
    side and holes on the p side constitute
    electric current, the magnitude of
    which is given by $I = (V-V_b)/R$

Note, from now on in these diagrams, we shall refer voltages (i.e. potentials)
to the (-) or "earth" side of the applied emf.

4.8 Reverse bias

1. Free electrons on the n side are attracted to battery
   positive causing them to move away from the junction
   causing additional build up of positive charge near the
   junction. The depletion region widens.

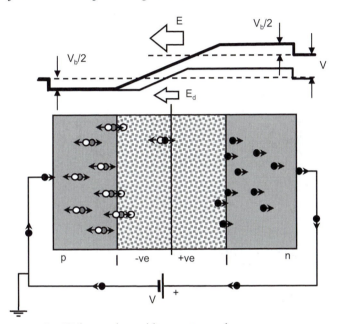

2. Holes on the p side are attracted
   towards the battery negative
   causing additional negative charge
   near the junction.

3. After a very short period, equilibrium is established
   where the attraction of mobile carriers to the
   externally applied voltage source is balanced by a
   build-up of charge on either side of the junction.
   Movement of majority carriers ceases, no current
   flows.

   Thermally generated minority carriers in the
   depletion layer may be swept cross junction
   by the field. This is called **leakage current**.

4.9 Potential diagrams

Forward bias

After being given an initial boost uphill by the voltage source V, electrons (now with potential energy) have a downhill run all the way around the circuit. Potential energy is lost after going through the resistor (heat), and then again through the contact potentials (overcoming diffusion "force").

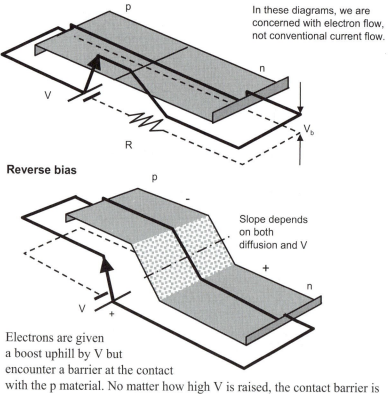

In these diagrams, we are concerned with electron flow, not conventional current flow.

Reverse bias

Slope depends on both diffusion and V

Electrons are given a boost uphill by V but encounter a barrier at the contact with the p material. No matter how high V is raised, the contact barrier is raised along with it. Since there is never any downhill path to the + terminal of V, no current flows.

If an electron appears on the slope (i.e. thermally generated within the depletion region), then the electron immediately rolls downhill to the n side acquiring sufficient "kinetic energy" to overcome the contact barrier and thus be transported around to the emf and given a boost uphill. This is reverse bias leakage current.

4.10 Diode

A p-n junction will conduct current in forward bias and act as an open
circuit in reverse bias. Such an action is called **rectification** and the device
as a whole is called a **diode**.

A perfect diode
would present
zero resistance in
forward bias and
infinite resistance
in reverse bias.

A simple model of
a diode includes
the potential
barrier V_b.

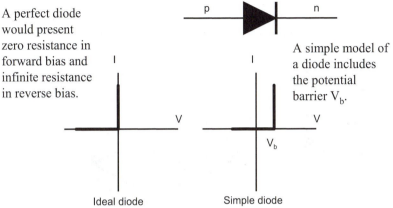

Ideal diode   Simple diode

A real diode has a slight forward resistance,
the potential barrier, leakage current and a
reverse breakdown voltage.

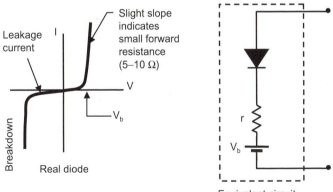

Equivalent circuit

4.11 Diode equation

Maxwell-Boltzmann statistics applied to the diffusion of charge carriers can predict current density across the junction in forward and reverse bias. The resulting equation is:

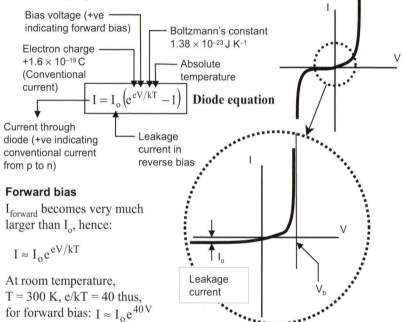

Bias voltage (+ve indicating forward bias)

Electron charge +1.6 × 10⁻¹⁹ C (Conventional current)

Boltzmann's constant 1.38 × 10⁻²³ J K⁻¹

Absolute temperature

$$I = I_0 \left(e^{eV/kT} - 1 \right)$$ **Diode equation**

Current through diode (+ve indicating conventional current from p to n)

Leakage current in reverse bias

Forward bias

$I_{forward}$ becomes very much larger than I_o, hence:

$$I \approx I_o e^{eV/kT}$$

At room temperature, T = 300 K, e/kT = 40 thus, for forward bias: $I \approx I_o e^{40\,V}$

Leakage current

For a linear resistor, V/I = R, but here, the relationship between V and I is not a constant, but is exponential. Hence, the slope of the line at any point gives the "resistance" of the forward bias junction.

$$\frac{dI}{dV} = 40\,I_o e^{40\,V}$$

$$= 40\,I$$

Hence,

$$\frac{dV}{dI} = \frac{1}{40\,I}$$

$$= r$$

$$\boxed{r = \frac{25}{I}}$$ **Dynamic resistance** of forward biased junction.

when I is expressed in mA

Reverse bias

V is negative and hence the exponential term is very small, thus:

$$I \approx -I_o$$

Leakage current is typically a few μA. Note, the diode equation says nothing about the possibility of breakdown.

4.12 Reverse bias breakdown

As the magnitude of the reverse bias voltage is increased, the current remains at I_o but eventually, the reverse bias field is so strong that thermally generated electrons (or holes) acquire enough kinetic energy to ionise atoms within the crystal structure. These in turn ionise other atoms leading to a very swift multiplication effect and a large current. This is called **avalanche breakdown**.

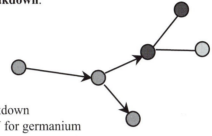

The reverse bias breakdown voltage is about 500 V for germanium and about 1 kV for silicon.

If the impurity doping density is high enough, then the depletion region is narrow enough (even in reverse bias) to allow the electric field across the region to be very high.The high accelerating field and narrow depletion region allows electrons to tunnel through. This is called **zener breakdown**. Zener diodes are designed to breakdown in reverse bias. They can withstand a relatively large reverse current without damage. The reverse bias voltage leading to zener breakdown is adjustable during manufacture of the device.

Typical **zener diodes** have breakdown voltages anywhere between 2 to 200 V depending on the application.

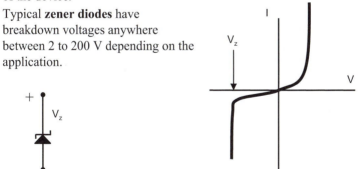

4.13 Rectification

One very common application of diodes is in **rectification** of an AC signal. That is, the conversion of AC into DC. In many cases, mains AC voltage has to be converted into a low DC voltage. The conversion from high voltage to low voltage is usually accomplished by a **transformer**, the output of which is a low voltage AC signal. This then has to be converted to a stable, DC output.

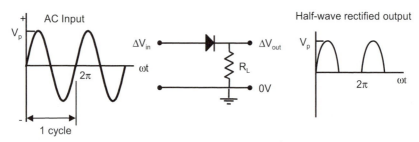

Full-wave rectification involves a clever arrangement of diodes to produce a DC signal but with a large ripple. This may be smoothed to give a fairly steady DC signal using various methods.

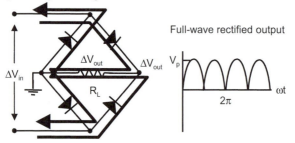

A steady "DC" output can be obtained by filtering the full-wave rectified signal with a capacitor.

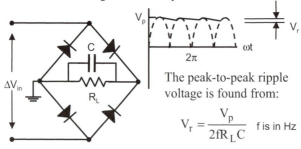

The peak-to-peak ripple voltage is found from:

$$V_r = \frac{V_p}{2fR_LC} \quad \text{f is in Hz}$$

4.14 Regulation

Zener diodes find special application as voltage regulators. They have a very sharp reverse bias breakdown characteristic. In a **voltage regulator**, the supply voltage can change significantly but the zener diode voltage V_Z does not change.

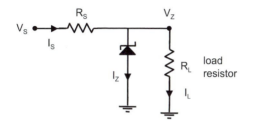

$V_L = V_Z$ and I_S is thus fixed and independent of R_L. If R_L increases, the zener passes more current to keep $V_L = V_Z$. When R_L is infinite, $I_Z = I_S$. Care must be taken to ensure that the maximum current through the zener diode does not cause overheating. Maximum current in the zener occurs at open circuit conditions (no R_L connected).

Example:
What resistor R_s is required to limit the power dissipation in the 4.8 V zener diode shown in the diagram above to 25 mW if V_s is 10 V?

Solution:

$$25\,\text{mW} = 4.8\text{I}$$

$$\text{I} = 5.2\,\text{mA}$$

$$10 - 4.8 = 5.2 \times 10^{-3} R_s$$

$$R_s = 1\,\text{k}\Omega$$

4.15 Clipper

A diode clipping circuit is useful for signal shaping. For example, in a geiger counter, a click is heard when an ionizing particle strikes the detector. Now, the detector itself does not produce voltage pulses of equal magnitude, but instead, an irregularly shaped signal. For the measurement of radiation, we simply require the rate at which pulses are produced and we are not concerned with the amplitude or shape of the pulse. A diode clipper can be used to produce a squared-up waveform which can then be fed into an audio amplifier or digital counter. The clipper ensures that the pulses fed to the counting circuit are uniform and that the counting circuit needs to only display the count rate.

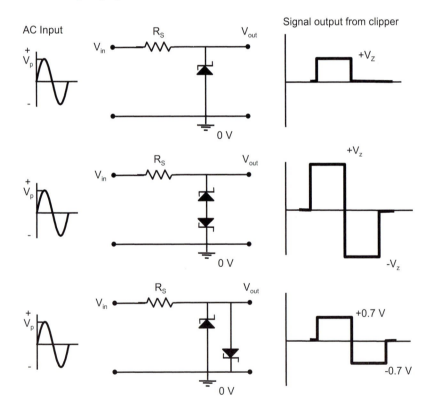

4.16 Clamp

The diode clamp is used for changing the reference voltage of an AC input signal.

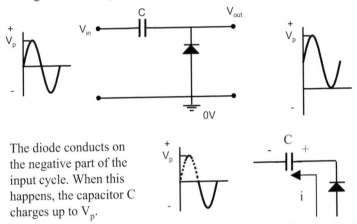

The diode conducts on the negative part of the input cycle. When this happens, the capacitor C charges up to V_p.

The potential of the right hand side of the capacitor with respect to the left hand side is $+V_p$. That is, when ΔV_{in} is at $-V_p$, the potential of the right hand side of the capacitor is 0 V. When the input ΔV_{in} reaches 0 V, $V_{out} = +V_p$ since the capacitor remains charged - the diode does not allow the capacitor to discharge.

On the positive half cycle, the left hand side of the capacitor is brought to a potential of $+V_p$. The right hand side of the capacitor must now be at a potential of $+2V_p$ with respect to 0 V.

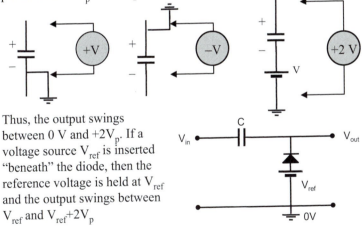

Thus, the output swings between 0 V and $+2V_p$. If a voltage source V_{ref} is inserted "beneath" the diode, then the reference voltage is held at V_{ref} and the output swings between V_{ref} and $V_{ref}+2V_p$

4.17 Pump

A diode pump is a circuit which produces a steady DC signal whose magnitude is proportional to the rate of arrival of voltage pulses at the input. We have seen how a clamp may be used to square up irregularly shaped pulses, now we shall see how to convert the train of pulses into a "rate" (i.e an output voltage which is proportional to the number of pulses per second).

The diode pump is how a tachometer works. Pulses from the ignition system are received and converted into a steady DC voltage whose magnitude thus depends on engine rpm. The DC voltage then drives a moving coil meter which is calibrated to read rpm.

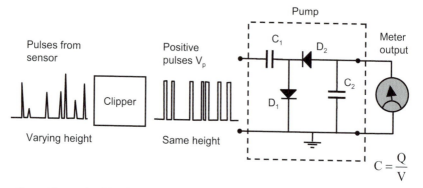

$$C = \frac{Q}{V}$$

A positive pulse V_p charges capacitor C_1 via D_1. The value of C is such that it can charge fully during the positive value of the input pulse (small time constant). Thus, a single pulse fully charges C_1. The right hand side of C_1 is negative w.r.t. the left hand side. When the left hand side returns to 0 V, the right hand side of C_1 is negative w.r.t. the top side of C_2. The negative charge is now distributed between C_1 and C_2 since D_2 is now conducting. Most of the charge on C_1 is transferred to C_2 (capacitors in parallel - voltage across each is the same, charge on each depends on C and here, $C_2 >> C_1$).

Now, if another pulse V_p arrives at the input, D_1 turns on as C_1 charges up and D_2 turns off. This leaves most of the original -ve charge on the "top" plate of C_2. C_1 accepts more charge from the input pulse and this is again transferred to C_2 on the next fall to 0 V at the input. The current through the meter depends upon the amount of accumulated charge on C_2. The faster the rate of input pulse, the greater the accumulation of charge on C_2. In order to indicate pulse rate, it is important that the peak voltage of the input pulses are all the same. The shape or width of them then does not matter.

4.18 Photodiode

A photodiode employs the **photovoltaic** effect to produce an electric current which is a measure of the intensity of incident radiation.

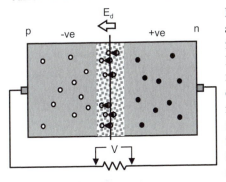

Diffusion of electrons and holes across the junction leads to the formation of a **barrier potential** leading to field E_d and a **depletion region**. When a photon creates an **electron-hole pair** in the depletion region, the resulting free electron is swept across the junction towards the n side (opposite direction of E_d). Current will flow in external circuit as long as photons of sufficient energy strike the material in the depletion region.

Even though the photodiode generates a signal in the absence of any external power supply, it is usually operated with a small reverse bias voltage. The incident photons thus cause an increase in the **reverse bias leakage current** I_o.

The reverse bias leakage current is directly proportional to the luminous intensity. Responsivity is in the order of 0.5 A W^{-1}.

Avalanche photodiodes operate in reverse bias at a voltage near to the break-down voltage. Thus, a large number of electron-hole pairs are produced for one incident photon in the depletion region (internal ionisation).

Phototransistors provide current amplification within the structure of the device. Incident light is caused to fall upon the reverse-biased collector-base junction. The base is usually not connected externally and thus the devices usually only have two pins. Increasing the light level is the same as increasing the base current in a normal transistor.

Schottky photodiodes use electrons freed by incident light at a metal-semiconductor junction. A thin film is evaporated onto a semiconductor substrate. The action is similar to a normal photodiode but the metal film used may be constructed so as to respond to short wavelength blue or ultraviolet light only since only relatively high energy photons can penetrate the metal film and affect the junction.

PIN photodiode is a p-n junction with a narrow region of intrinsic semiconductor sandwiched between the p- and n-type material. This insertion widens the depletion layer thus reducing the junction capacitance and the time constant of the device - important for digital signal transmission via optical cable.

4.19 LED

The light-emitting diode (LED) operates in forward bias and generates a photon when electrons and holes recombine near the junction.

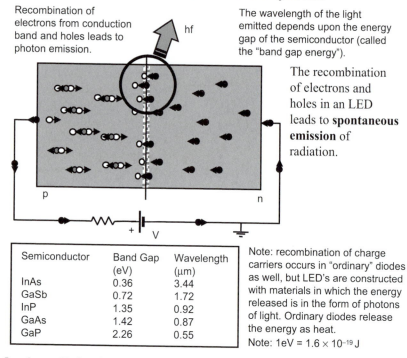

Recombination of electrons from conduction band and holes leads to photon emission.

hf

The wavelength of the light emitted depends upon the energy gap of the semiconductor (called the "band gap energy").

The recombination of electrons and holes in an LED leads to **spontaneous emission** of radiation.

p

n

Semiconductor	Band Gap (eV)	Wavelength (µm)
InAs	0.36	3.44
GaSb	0.72	1.72
InP	1.35	0.92
GaAs	1.42	0.87
GaP	2.26	0.55

Note: recombination of charge carriers occurs in "ordinary" diodes as well, but LED's are constructed with materials in which the energy released is in the form of photons of light. Ordinary diodes release the energy as heat.
Note: $1\,eV = 1.6 \times 10^{-19}\,J$

In a **laser diode**, photons arising from spontaneous emission are reflected back and forth between the polished faces of the device. These photons are absorbed within the crystal releasing an electron into the conduction band. However, simultaneously with this absorption, electrons in the conduction band also fall back into the valence band and a photon of the same frequency is emitted. This is **stimulated emission**.

Absorption and stimulated emission occur simultaneously and with equal probability. However, in a laser diode, the geometry of the mirrored faces and the doping of the crystal ensures that when the current through the device is sufficient, there are more electrons in the conduction band than in the valence band (**population inversion**) and the stimulated emission of photons has a greater chance of occurring than absorption. The emitted photons all have the same phase and frequency and are emitted as laser light out through one of the partially mirrored sides of the device.

Review questions

1. Briefly describe the essential features of an electrical insulator, semi-conductor and a conductor.

2. Explain the origin of leakage current in a reverse-biased p-n junction.

3. It can be shown that the barrier potential of a p-n junction can be calculated from:

$$V_B = \frac{kT}{e} \ln\left(\frac{N_A N_D}{n_i^2} \right)$$

Assuming that for germanium the N_A is 5×10^{16} cm^{-3} and N_D is 1×10^{18} cm^{-3}, calculate the value of the barrier potential V_B at T = 300 K. Assume $n_i = 1 \times 10^{15}$cm^{-3}.

(Ans: 0.28 V)

4. Calculate the apparent resistance of a forward biased p-n junction at room temperature (300 K) when the current through the junction is 5 mA.

(Ans: 5 Ω)

5. In the circuit below, calculate the value of R_Z required so that the power dissipated by the 4.8 V zener diode does not exceed 25 mW.

(Ans: 1 kΩ)

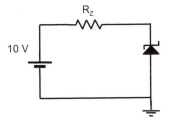

6. In a logic gate which implements the logic "OR" function, a steady 5 V DC signal is produced at the output when either one of 3 inputs is at 5 V. Implement this circuit using diodes.

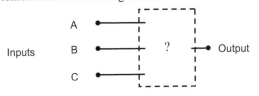

7. A motor vehicle battery charger consists of a step-down
   transformer, a resistor and a diode. The transformer supplies a
   voltage ΔV_s to the diode and resistor as shown.

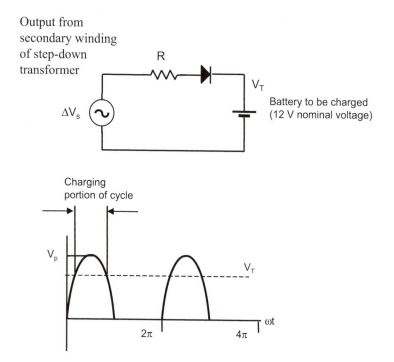

The battery voltage V_T rises as charging progresses. The peak charging
voltage is 14 V. At low charge conditions, (V_T is low), we require
maximum charging current (6 amps rms).

(a) Determine a value for the resistance R. Hint: when the battery is
    completely flat, the charging current is a maximum, further, the rms
    current for a half sine wave is $I_p/2$.
(b) Determine the maximum reverse voltage applied to the diode;
(c) Calculate the peak value of the charging current when the battery voltage
    reaches 12 V;
(d) Determine the rms current when the battery terminal voltage reaches
    12 V (Hint, this is NOT $I_p/2$).

(Ans: 1.17 Ω, 26 V, 1.7 A, 0.67 A)

5. Bipolar junction transistor

Summary

$$\frac{I_c}{I_b} = h_{fe} \quad \text{Current gain}$$

$$I_c = -\frac{1}{R_c} V_{ce} + \frac{V_{cc}}{R_c} \quad \text{Load line (simple bias)}$$

$$V_b = I_b R_b + V_{be} \quad \text{Simple bias}$$

$$V_{cc} = I_c R_c + V_{ce}$$

$$I_c = -\frac{1}{R_c + R_e} V_{ce} + \frac{V_{cc}}{R_c + R_e} \quad \text{Load line (emitter bias)}$$

$$\frac{\Delta V_{out}}{\Delta V_{in}} = A_v \quad \text{Voltage gain}$$

5.1 Bipolar junction transistor - construction

How does it work?

Step 1. Start with a reverse-biased p-n junction

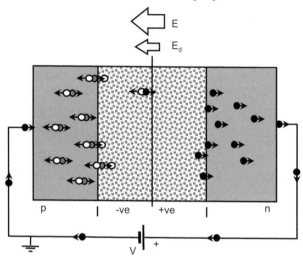

Note: the field $E+E_d$ only exists across the depletion region. Thus, once equilibrium is established, the electrical potential between the p and the n type material is equal and opposite to V. Only mobile charge carriers originating within the depletion region are accelerated by the field causing reverse bias leakage current to flow.

Step 2. Create an additional p-n junction on the left by adding some n-type material and make the new junction forward-biased by applying V_b

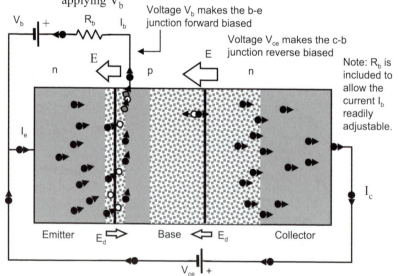

Voltage V_b makes the b-e junction forward biased

Voltage V_{ce} makes the c-b junction reverse biased

Note: R_b is included to allow the current I_b readily adjustable.

Here's what happens:

- Base-emitter junction is a forward-biased p-n junction so when the voltage $V_{be} > 0.7V$ (for silicon) then the junction becomes conducting (just like a diode).

- Electrons coming from the heavily doped emitter cross the junction but before they have a chance to combine with holes in the p-type base and travel to the V_b positive terminal, they get swept up by the strong field which exists around the collector base junction which is reverse-biased.

  - Because the base is made <u>lightly doped</u> (so that recombination in the base is unlikely to occur) and is made <u>very thin</u> (so that electrons coming across the forward-biased b-e junction do not have far to go before they "overshoot" and fall into the field across the c-b junction).

- Hence, only a few electrons go towards +ve V_b and most are attracted across the collector base junction and cause a large current in the collector. - If the electrons coming across into the base were not attracted across the c-b junction (e.g. if $V_{ce} = 0$) then they would simply go towards V_{bb} and there would be a large base current. But, because of the large field surrounding the reverse-biased c-b junction, these electrons are "siphoned off" thus causing $I_c \gg I_b$ and that any increase or decrease in I_b would be reflected by an increase or decrease in I_c.

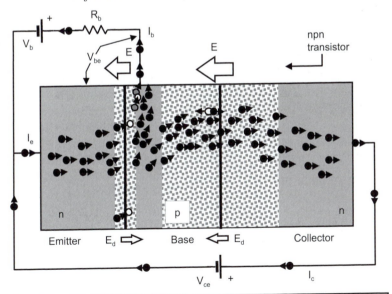

The result is a very small current in the base circuit, and a very large current in the collector circuit.

5.2 Bipolar junction transistor - operation

How is the **collector current** controlled?

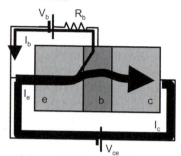

Electrons only cross the b-e junction
because it is forward biased by V_b.
Hence, all electrons which cross this
junction initially want to go to $+V_b$. If
the base current were to be increased,
(by increasing the magnitude of V_b),
then there would be more electrons
(per second) wanting to go towards $+V_b$

and hence more electrons (per second) being siphoned off towards the
collector by the reverse bias field at the c-b junction. The magnitude of the
base current controls the magnitude of the collector current. The ratio of the
two currents is called the **current gain** and given the symbol h_{fe}.

$$\frac{I_c}{I_b} = h_{fe}$$

Electrons which do go to the collector find their way around to the
emitter again joining those which went through the base. Thus, in
the emitter, there are two currents, I_b and I_c. Thus: $I_e = I_c + I_b$

If V_{ce} were to be turned off, then one might expect that all these electrons
would go towards $+V_b$ and the base current would increase quite dramatically.

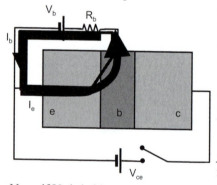

In practice, something else actually
happens. This is where the resistance
R_b comes into operation. R_b can be
deliberately put there by us as an
external resistor, or if no external
resistor, may represent the internal
resistance of the voltage source V_b.
Now, we know that the voltage V_b
supplies a voltage drop across the b-e
junction of about 0.7 V and the
remainder appears across R_b.

Now, if V_b is held constant, but I_b were to increase due to the removal of the
reverse bias field at c-b junction, then there would be an increase in the
voltage drop across R_b. But, since V_b is a constant, this must mean that there is
a reduction in the voltage across the forward bias b-e junction. Now, if V_{be} is
decreased, then due to the exponential nature of the IV characteristics of the
p-n junction, there is a substantial reduction in I_b. But this reduction in I_b also
reduces the voltage drop across R_b hence tending to increase V_{be} - negative
feedback. The net result is a new equilibrium with V_{be} reduced slightly and a
less-than-expected I_b is observed.

The **base current** is controlled by a resistor R_b which is inserted between the voltage supply V_b and the base so that I_b may be adjusted by adjusting V_b. Thus, consider the circuit below which shows the "base-emitter" half of the transistor:

In this circuit, V_{be} remains at somewhere near 0.7 V and increasing the voltage V_b simply increases the voltage drop across R_b due to the increased current I_b. (A large change in I_b does not change V_{be} all that much compared to the change in voltage across R_b) Thus, with the resistor R_b in place, we may adjust I_b by adjusting V_b

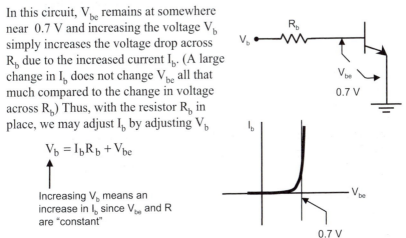

$$V_b = I_b R_b + V_{be}$$

Increasing V_b means an increase in I_b since V_{be} and R are "constant"

A BJT is a current controlled device. Resistors may be used to convert current control into voltage control, but the essential feature is that for a given base current I_b, there is a fixed value of collector current I_c.

A plot of I_c vs I_b shows that the **current gain** h_{fe} is fairly constant. In practice, h_{fe} depends on manufacturing variables and is very different between actual transistor components. Values between 100–300 are typical.

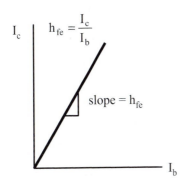

$$h_{fe} = \frac{I_c}{I_b}$$

slope $= h_{fe}$

5.3 Transistor characteristic

Transistor action can be summarised in one figure which is
called the **transistor characteristic**.

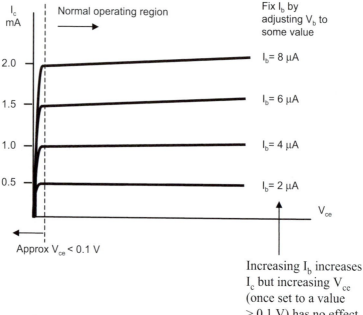

Increasing I_b increases
I_c but increasing V_{ce}
(once set to a value
> 0.1 V) has no effect.
The transistor is a
current controlling
device. I_c is constant
for a given value of I_b

Note: in practice there is some slight increase in I_c when V_{ce} is increased. As V_{ce} is
increased, the width of the depletion region associated with the reverse-biased c-b
junction increases and the reverse bias field also increases. When this happens, for a
given base current I_b, a greater proportion of the electrons are "collected" by the
collector and I_b reduces. However, a reduction in I_b leads to a reduction in the voltage
across the base resistor and an increase in voltage across the b-e forward bias junction.
This leads to an increase in I_b due to the feedback mechanism discussed previously.
The overall effect is for a fairly constant I_b and a slightly increasing I_c with increasing V_{ce}

5.4 Load line

Consider this circuit:

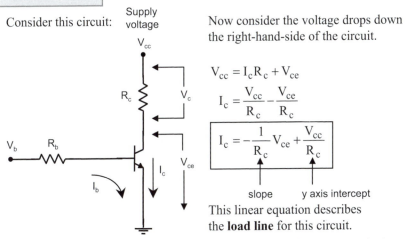

Now consider the voltage drops down the right-hand-side of the circuit.

$$V_{cc} = I_c R_c + V_{ce}$$

$$I_c = \frac{V_{cc}}{R_c} - \frac{V_{ce}}{R_c}$$

$$\boxed{I_c = -\frac{1}{R_c} V_{ce} + \frac{V_{cc}}{R_c}}$$

        slope   y axis intercept

This linear equation describes the **load line** for this circuit.

V_{ce} here is determined by the *circuit*. That is, I_b controls I_c, which results in a voltage drop across R_c. Hence, if I_b goes up, I_c goes up and voltage V_c goes up and V_{ce} goes down.

The load line may be superimposed on the transistor characteristic.

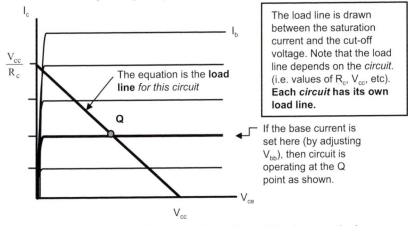

The load line is drawn between the saturation current and the cut-off voltage. Note that the load line depends on the *circuit*. (i.e. values of R_c, V_{cc}, etc). **Each *circuit* has its own load line.**

The equation is the **load line** *for this circuit*

If the base current is set here (by adjusting V_{bb}), then circuit is operating at the Q point as shown.

The load line shows the allowable values of I_c and V_{ce} for a particular circuit. The Q point is the value of I_c and V_{ce} which might be measured for a circuit at some particular value of V_b. The corresponding base current may be obtained from the transistor characteristic curve which is coincident with the Q point.

5.5 Saturation

1. At low values of V_{ce} (i.e. high values of I_b) circuit follows the load line *until the transistor saturates.*

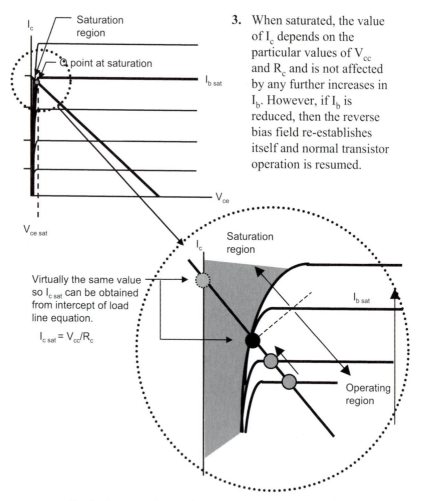

3. When saturated, the value of I_c depends on the particular values of V_{cc} and R_c and is not affected by any further increases in I_b. However, if I_b is reduced, then the reverse bias field re-establishes itself and normal transistor operation is resumed.

Virtually the same value so $I_{c \, sat}$ can be obtained from intercept of load line equation.

$I_{c \, sat} = V_{cc}/R_c$

2. In the operating region, increasing I_b results in an increase in I_c and operating point moves up the load line. Eventually, a point is reached $I_{b \, sat}$ where an increase in I_b results in no further increase in I_c. i.e. characteristic curves for $I_b > I_{b \, sat}$ all intersect load line at the same place $I_{c \, sat}$

5.6 Transistor switch

When the base current I_b is zero, no collector current I_c will flow (no matter how high V_{ce} might be) since there are no charge carriers present to be swept across the reverse-biased collector-base junction.

1. Transistor is OFF

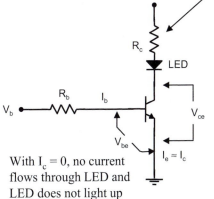

In this circuit, a resistor R_c has been inserted between the supply voltage and the collector. This resistor limits the value of the collector current I_c to some maximum so as to not overload the LED.

If V_{bb} is reduced to zero, then I_b goes to zero and so does I_c.

$$I_c = h_{fe}I_b$$

A transistor is a current controlled device.

With $I_c = 0$, no current flows through LED and LED does not light up

2. Transistor is ON

Increasing V_b increases I_b and hence allowing greater I_c to pass through the collector to the emitter.

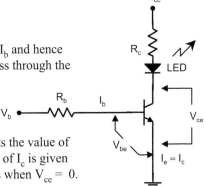

The presence of R_c limits the value of I_c . The maximum value of I_c is given by $V_{cc} = I_c R_c$ and occurs when $V_{ce} = 0$.

In this circuit, the voltage V_{ce} is not a constant but depends on the current I_c. If I_c increases, then so does the voltage drop across R_c and thus voltage V_{ce} decreases. Thus, if I_b is increased such that I_c becomes equal to the maximum allowed by R_c, then V_{ce} must be approaching zero volts - this is called **saturation**.

5.7 Simple bias

Bias voltage is a steady dc voltage applied to the transistor at all times. This voltage is necessary so that the p-n junctions are placed in forward and reverse bias as appropriate so that correct transistor operation is obtained.

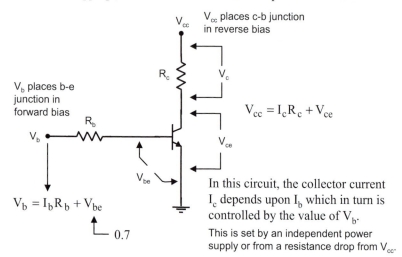

V_{cc} places c-b junction in reverse bias

V_b places b-e junction in forward bias

$$V_{cc} = I_c R_c + V_{ce}$$

$$V_b = I_b R_b + V_{be}$$

In this circuit, the collector current I_c depends upon I_b which in turn is controlled by the value of V_b. This is set by an independent power supply or from a resistance drop from V_{cc}.

If h_{fe} increases (e.g. by changing the transistor for another one) then the transistor characteristic is changed and for a given I_b (fixed by V_b), I_c increases and the relative position of the operating point (the Q point) on the load line changes.

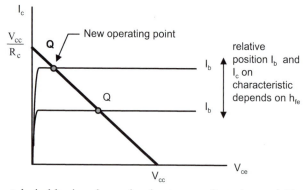

New operating point

relative position I_b and I_c on characteristic depends on h_{fe}

This is not desirable since h_{fe} varies due to manufacturing variables and may vary anywhere from between 100–200. For reasons to be given later, we require the operating point to remain approximately in the middle of the load line even when h_{fe} changes.

5.8 Emitter bias

The disadvantages of the simple bias arrangement can be overcome by the **emitter bias** circuit:

Remove R_b and let V_b represent the voltage measured at the base. Then, insert R_e between the emitter and earth.

Consider the "input" side of this circuit;

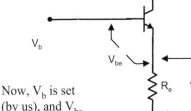

Now, V_b is set (by us), and V_{be} remains at about 0.7 V, thus the voltage drop across R_e is given by: $V_e = V_b - V_{be}$

The potential at the top of R_e is fixed by setting V_b. However, there are two currents running through R_e, i.e. $I_e = I_b + I_c$ but for all practical purposes: $I_e \approx I_c$

Thus: $V_e = I_c R_e$

$$I_c R_e = V_b - V_{be}$$

$$I_c = \frac{V_b - V_{be}}{R_e} \quad \text{all independent of } h_{fe}$$

That is, I_c is fixed by the value of R_e But, you say, what about I_b? Doesn't I_b control I_c? Yes it does, but here the focus is on I_c and I_b follows (according to the value of h_{fe}). That is, we choose a "design value" of I_c by selecting R_e and V_b (assuming $V_{be} = 0.7$). The "correct" base current $I_b (= I_c/h_{fe})$ automatically flows when V_b is applied.

The **load line** for the circuit becomes:

$$I_c = -\frac{1}{R_c + R_e} V_{ce} + \frac{V_{cc}}{R_c + R_e}$$

The underlying assumption is that h_{fe} is large. If I_b ($= I_c/h_{fe}$) is included, then:

$$V_e = \left(I_c + \frac{I_c}{h_{fe}} \right) R_e$$

$$I_c + \frac{I_c}{h_{fe}} = \frac{V_{bb} - V_{be}}{R_e}$$

5.9 Stabilisation

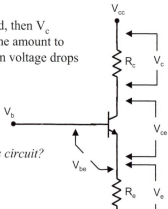

In the emitter bias circuit, if R_c is decreased, then V_c decreases and V_{ce} must increase by the same amount to compensate. I_c remains constant to maintain voltage drops from V_b down to earth.

If R_e increases, then V_e increases. Any increase in V_e has an important effect on I_b as will be described below.

What happens if h_{fe} changes in emitter bias circuit?

Coming down the right-hand-side of the circuit, we have:

$$V_{cc} = V_c + V_{ce} + V_e$$
$$= I_c R_c + V_{ce} + I_c R_e$$
$$= I_c (R_c + R_e) + V_{ce}$$

If h_{fe} goes up (e.g. transistor is heated), then, for a given I_b, I_c increases by some small amount ΔI_c. If I_c increases then V_c also increases and so also does V_e. Thus, V_{ce} must decrease to keep the voltage drops from V_{cc} to earth consistent. But: $V_b = V_{be} + V_e$

If V_e increases, then V_{be} must decrease since V_b is set: -

- but V_{be} is the forward biasing voltage for the base-emitter junction. Hence a small drop in V_{be} results in a large drop in I_b

- but a decrease in I_b results in a decrease in I_c. A new equilibrium is reached and I_c settles down to its former value with a reduced I_b being the result of the increased h_{fe}. $I_c/I_b = h_{fe}$ as always but in this circuit, I_c is controlled (and hence the Q point) and V_{be} and hence I_b changes to account for any variations in h_{fe} The circuit is stabilised against changes in h_{fe}.

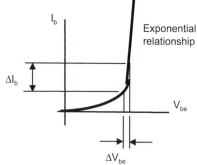

5.10 Voltage amplifier

Although we have so far examined the
bias characteristics of the emitter bias
circuit, we may note that the voltage at the
collector $V_{ce}+V_e$ could be considered an
output voltage V_{out} and V_b may be
considered an input voltage.

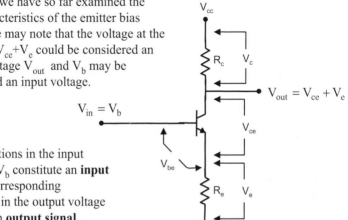

$$V_{in} = V_b$$

$$V_{out} = V_{ce} + V_e$$

Any variations in the input
voltage ΔV_b constitute an **input
signal**. Corresponding
variations in the output voltage
ΔV_{out} is an **output signal**.

Small variations ΔV_{in} cause large variations ΔV_{out} because of the effect of the
current gain h_{fe}. The ratio of the two signals is the **voltage gain** A_v

$$\boxed{\frac{\Delta V_{out}}{\Delta V_{in}} = A_v}$$

Later we will see how this voltage gain is obtained in more detail. For now,
let us assume that such a voltage gain is possible and that small variations
at the input ΔV_{in} lead to large variations on the output ΔV_{out}

But, a varying input signal
must not send the transistor to
cut-off or **saturation** since
then the output signal will be
distorted or clipped. Further,
the varying input signal must
be always positive (i.e. $V_{be} >$
0.7 volts) so that the base-
emitter junction is always in
forward bias.

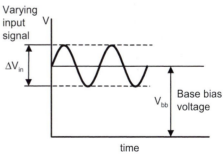

A bias voltage is a steady DC voltage applied to the transistor base so that
correct transistor operation is obtained without cut-off or saturation when a
varying input signal ΔV_{in} appears on the input.

5.11 Bias

The DC bias is set so that the operating point (the Q point) is halfway along the load line. This ensures that we get maximum output voltage swing at V_{out} without transistor saturating or reaching cut-off.

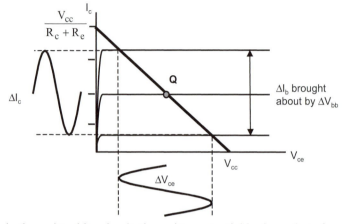

Since, in the emitter bias circuit, the emitter potential is above that of earth (V_e) then at saturation, $V_{out} = V_e$. At cut-off, $V_{out} = V_{cc}$, thus the output voltage can only swing between V_{cc} and V_e *in this circuit.*

We can now call the voltage at the base "V_{bb}" (for **base bias voltage**). A convenient way of supplying a steady V_{bb} is to use a voltage divider from the supply V_{cc}.

$$I_c = \frac{V_{bb} - V_{be}}{R_e}$$

We must choose the bias voltage V_{bb} so that I_c falls in the middle of the load line.

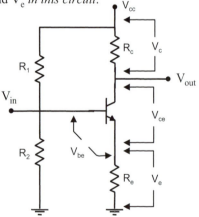

The supply voltage V_{cc} provides both power for the output signal and also a voltage V_{ce} which keeps the collector base junction in reverse bias. The combined value of R_c and R_e limit the maximum collector current (which may need limiting so as to not overload the transistor) and R_e should not be made too large so that there is sufficient allowable voltage swing at the output V_{out}.

Review questions

1. Determine the current gain (h_{fe}) of the transistor characteristic
   shown below in the normal operating region

(Ans: $h_{fe} = 400$)

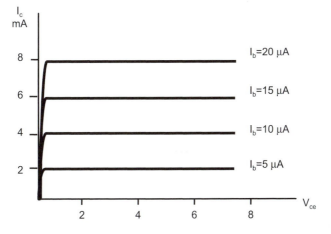

2. Using the equation below, plot I_c vs V_{ce} for $I_0 = 10^{-15}$A and $V_A = 100$ V for
   $V_{ce} = 5, 10, 15, 20, 25$ V and $V_{be} = 0.65, 0.675$ and 0.7V at $T = 300$ K
   (leave plenty of room on the negative x axis)

$$I_c = I_0 e^{\frac{qV_{be}}{kT}} \left(1 + \frac{V_{ce}}{V_A}\right)$$

$k = 1.38 \times 10^{-23}$
$q = 1.6 \times 10^{-19}$

V_A is called the **Early voltage** and is the point on the $-$ve V_{ce} axis
where the characteristic curves meet. Draw these characteristic curves
on your graph and extrapolate them to the V_{ce} axis and indicate the
Early voltage.

3. The switching circuit shown below converts a 0 to 6 V pulse into a 10 V
   to 5 V pulse. Determine the values of the components R_b, R_e and R_c
   given that the current gain is 50, and the maximum current drawn from
   the 10 V power supply is 10 mA. Assume $V_{be} = 0.7$ V

(Ans: R_c, R_e 500 Ω, R_b 1.5 kΩ)

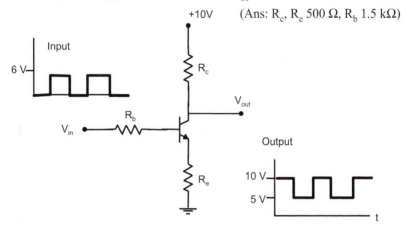

4. The transistor switch below is operated by a 4 volt signal.
   (a) Calculate the corresponding output voltages for input voltages of both 0
       and 4 volts. Assume that the transistor is put either into saturation or
       cut-off by the input signal;
   (b) Calculate a suitable value of R_b.

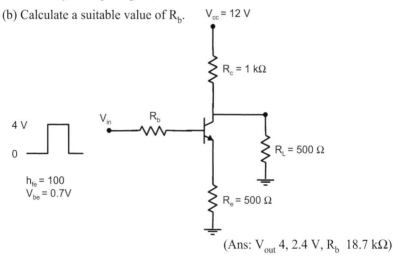

(Ans: V_{out} 4, 2.4 V, R_b 18.7 kΩ)

6. Common emitter amplifier

Summary

Common emitter amplifier

$$V_T = I_b R_T + V_{be} + I_c R_e$$
$$= I_b R_T + V_{be} + h_{fe} I_b R_e$$
$$= I_b (R_T + h_{fe} R_e) + V_{be}$$

$$h_{fe} = \frac{I_c}{I_b}$$

$$V_c = I_c R_c$$

$$I_c = \frac{V_e}{R_e}$$

$$V_{cc} = V_c + V_{ce} + V_e$$

$$V_{bb} = V_{be} + V_e$$

$$h_{ie} = \frac{25}{I_c} h_{fe}$$

$$A_v = -\frac{R_{out}}{h_{ie}} h_{fe}$$

$$= -\frac{R_{out}}{r_e}$$

$$= -R_{out} \frac{I_c}{25}$$

6.1 Coupling capacitors

A voltage amplifier produces an output voltage that is proportional to the input voltage. Most signals that require magnification are AC.

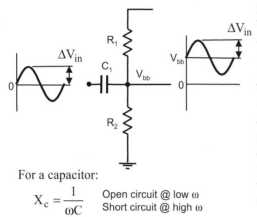

R_1 and R_2 provide the proper bias voltage to the base emitter junction. The input signal ΔV_{in} needs now to be superimposed onto the dc bias level but the signal source should not load the circuit in any way. An isolating input **coupling capacitor** C_1 is inserted to isolate transistor bias voltages from DC voltages from the signal source.

For a capacitor:

$$X_c = \frac{1}{\omega C}$$

Open circuit @ low ω
Short circuit @ high ω

Low frequency response is limited by C_1. The 3 dB point is calculated from $R\omega C = 1$ where R is the input resistance R_{in} of the circuit.

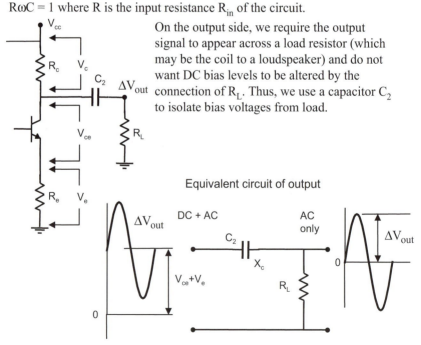

On the output side, we require the output signal to appear across a load resistor (which may be the coil to a loudspeaker) and do not want DC bias levels to be altered by the connection of R_L. Thus, we use a capacitor C_2 to isolate bias voltages from load.

Equivalent circuit of output

6.2 Bypass capacitor

In the stabilised biasing circuit, a resistor R_e was inserted to keep the
emitter potential a little bit above earth so that the collector current and V_{ce}
would not be affected by changes in h_{fe} (stable DC bias Q point).

Consider a slight increase in V_{bb} to
$V_{bb}+\Delta V_{bb}$. This would result in an
increase ΔI_b and hence increase ΔI_c.
But, since $\Delta I_c = h_{fe}\Delta I_b$, then the
resulting ΔV_e is larger than ΔV_{bb}.

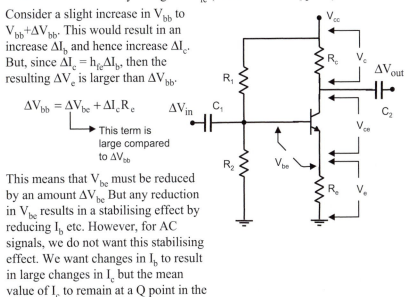

$$\Delta V_{bb} = \Delta V_{be} + \Delta I_c R_e$$

This term is
large compared
to ΔV_{bb}

This means that V_{be} must be reduced
by an amount ΔV_{be} But any reduction
in V_{be} results in a stabilising effect by
reducing I_b etc. However, for AC
signals, we do not want this stabilising
effect. We want changes in I_b to result
in large changes in I_c but the mean
value of I_c to remain at a Q point in the
middle of the load line.

Thus, on the one hand, we require R_e to provide DC bias stability but on
the other hand, we do not want R_e to reduce the signal, or AC voltage gain.
The solution? insert a "bypass" capacitor across R_e so that AC signals
proceed directly to earth while DC bias is still stabilised.

This bypass capacitor has a
significant effect on the AC
performance of the circuit and
we shall have cause to examine
its effect later on in more detail.

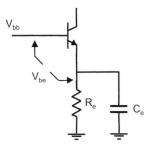

6.3 Voltage amplifier

Voltage amplifying circuit

- emitter bias circuit with voltage divider

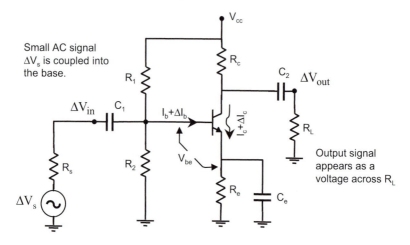

How does it work?

ΔV_{in} causes ΔV_{be}

ΔV_{be} results in ΔI_{b}

ΔI_{b} results in ΔI_{C}

ΔI_{c} results in ΔV_{ce} which is ΔV_{out}

Note: if ΔV_{in} is positive (an increase in voltage at V_{b}) then ΔV_{be} increases which results in an increase in I_{b} and hence an increase in I_{c}. Thus, V_{c} goes up and V_{ce} goes down by ΔV_{ce}. That is, the output is out of phase with the input signal by 180°.

How to analyse this circuit? ⟶ Treat AC and DC separately

Given resistances and supply voltage, what are all the other voltages?

DC	AC
• Open all capacitors (open circuit)	• Reduce DC sources to 0 volts (AC earth) • Short all capacitors

6.4 DC Analysis

Start with complete circuit:

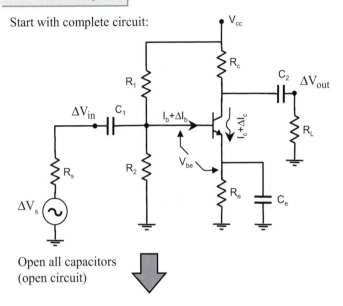

Open all capacitors
(open circuit)

DC circuit

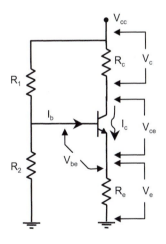

1. Begin with simplifying the DC bias circuit using Thevenin's theorem

h_{fe}, V_{cc}, R_1, R_2, R_c, R_e known

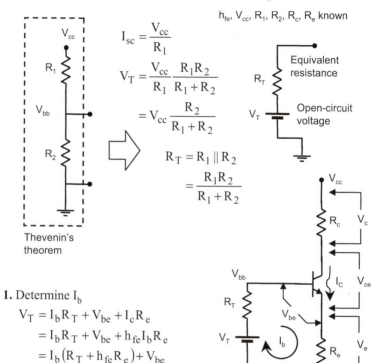

$$I_{sc} = \frac{V_{cc}}{R_1}$$

$$V_T = \frac{V_{cc}}{R_1} \frac{R_1 R_2}{R_1 + R_2}$$

$$= V_{cc} \frac{R_2}{R_1 + R_2}$$

$$R_T = R_1 \parallel R_2$$

$$= \frac{R_1 R_2}{R_1 + R_2}$$

R_T Equivalent resistance

V_T Open-circuit voltage

Thevenin's theorem

1. Determine I_b

$$V_T = I_b R_T + V_{be} + I_c R_e$$
$$= I_b R_T + V_{be} + h_{fe} I_b R_e$$
$$= I_b (R_T + h_{fe} R_e) + V_{be}$$

0.7

2. Determine I_c

$$h_{fe} = \frac{I_c}{I_b}$$

3. Determine $V_c = I_c R_c$

4. Determine V_e $I_c = \dfrac{V_e}{R_e}$

5. Determine V_{ce}

$$V_{cc} = V_c + V_{ce} + V_e$$

6. Determine

$$V_{bb} = V_{be} + V_e$$

Note: although it might seem objectionable that we require h_{fe} to begin this calculation sequence, this arises because we are taking into account the base current I_b and its effect on the current through R_2. If we disregard the base current, and assume that the current through R_1 and R_2 is much larger than I_b, then h_{fe} need not be known beforehand.

6.5 AC Analysis

Start with complete circuit:
- reduce DC sources to 0 volts
  (short circuit)
- short all capacitors

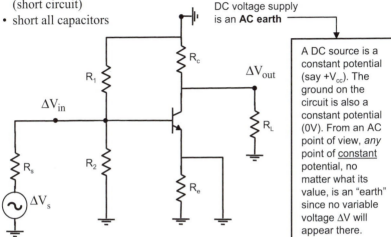

DC voltage supply
is an **AC earth**

A DC source is a constant potential (say $+V_{cc}$). The ground on the circuit is also a constant potential (0V). From an AC point of view, *any* point of <u>constant</u> potential, no matter what its value, is an "earth" since no variable voltage ΔV will appear there.

AC circuit

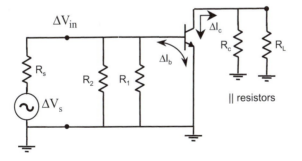

|| resistors

ΔV_{in} at the base produces a ΔV_{be} at the forward bias base-emitter junction (since the bypass capacitor shorts the emitter to ground). This ΔV_{be} results in a change in I_b by ΔI_b.

The resulting ΔI_b thus causes a ΔI_c (as per h_{fe}). This in turn creates a change in the voltage drop across R_c and thus also a change in V_{ce}

But $\Delta V_{ce} = \Delta V_{out}$, thus the question is, how can ΔV_{out} be calculated in terms of ΔV_{in}?

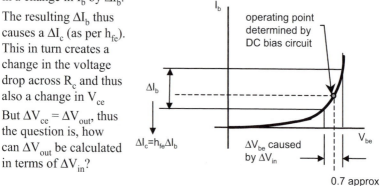

The first step is to determine: *"What base current ΔI_b is produced by the input voltage signal ΔV_{in}?"*

Once we know ΔI_b we can get ΔI_c and hence ΔV_{out}

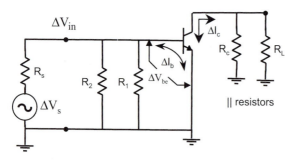

Observe that the presence of the bypass capacitor causes a portion (ΔI_b) of the AC input current (ΔI_s) to pass straight from ΔV_{in} through the forward bias base emitter junction directly to earth (the remainder going through the parallel resistors $R_{1,2}$). Thus, $\Delta V_{in} = \Delta V_{be}$.

To get ΔI_b, we must find out the resistance of the base-emitter junction (*as viewed from the base*). This is given the symbol h_{ie}.

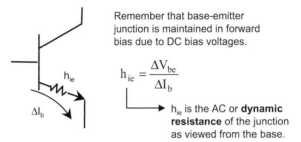

Remember that base-emitter junction is maintained in forward bias due to DC bias voltages.

$$h_{ie} = \frac{\Delta V_{be}}{\Delta I_b}$$

h_{ie} is the AC or **dynamic resistance** of the junction as viewed from the base.

How to get h_{ie}? It is given by the local slope of the I-V curve at the particular value of V_{be}, hence:

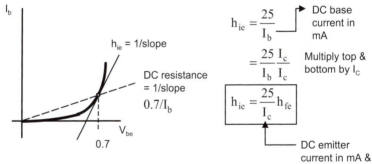

$h_{ie} = 1/\text{slope}$

DC resistance = 1/slope

$0.7/I_b$

$$h_{ie} = \frac{25}{I_b}$$ DC base current in mA

$$= \frac{25}{I_b}\frac{I_c}{I_c}$$ Multiply top & bottom by I_c

$$h_{ie} = \frac{25}{I_c}h_{fe}$$

DC emitter current in mA & letting $I_c = I_e$

From the emitter's point of view, a large current (ΔI_e) appears to pass across the b-e junction. But, from the base point of view, only a small current (ΔI_b) appears to pass across the b-e junction. Hence, when looking from the base to the emitter, the resistance h_{ie} appears large. When looking up from the emitter, the resistance appears to be small and is given the symbol r_e. The ratio of the two resistances is h_{fe}:

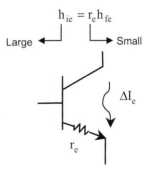

$$h_{ie} = r_e h_{fe}$$

Large   Small

ΔI_e

r_e

6.6 Input and output resistances - AC

Consider the resistances (more correctly "impedances") as seen from the input terminals of the circuit to earth.

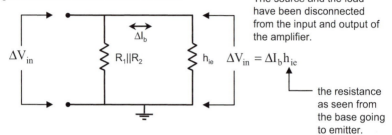

The source and the load have been disconnected from the input and output of the amplifier.

$\Delta V_{in} = \Delta I_b h_{ie}$

the resistance as seen from the base going to emitter.

The **input resistance** is $R_{in} = R_1 \| R_2 \| h_{ie}$

In a fully stabilised bias circuit, it is I_c which is "controlled" or designed and I_b adjusts according to the value of h_{fe}

It will be later shown that (for other reasons) $R_T = R_1 \| R_2$ is made $\gg h_{ie}$ so that the input resistance of the amplifier is dominated by the value of h_{ie} (since $h_{ie} \| R_T$). It is desirable to have a high input resistance so h_{ie} should be made to be fairly large - by choosing the lowest possible value of I_c.

For a given ΔI_b, there will be a constant ΔI_c in the collector, or output side of the circuit. That is, the transistor acts as a constant current source across which is connected a resistor R_c. This is the same situation as a Norton constant current source and parallel resistor. Thus, the output "half" of the amplifier circuit can be drawn:

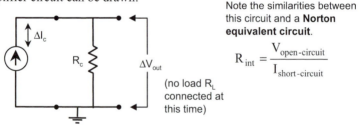

Note the similarities between this circuit and a **Norton equivalent circuit**.

$$R_{int} = \frac{V_{open-circuit}}{I_{short-circuit}}$$

(no load R_L connected at this time)

For the moment, we may refer to R_c as being the **output resistance** of the amplifier circuit R_{out}.

6.7 AC Voltage gain

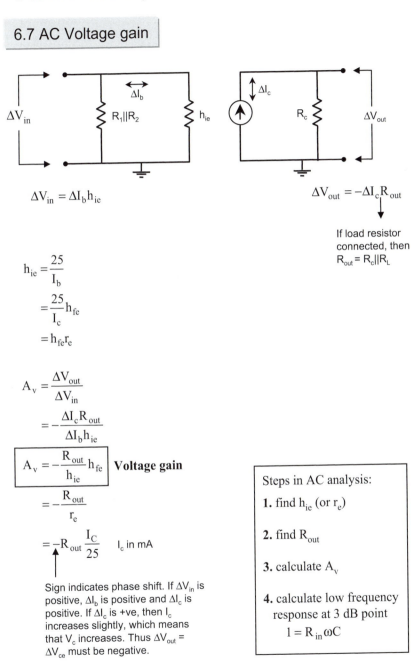

$$\Delta V_{in} = \Delta I_b h_{ie}$$

$$\Delta V_{out} = -\Delta I_c R_{out}$$

If load resistor connected, then
$R_{out} = R_c \| R_L$

$$h_{ie} = \frac{25}{I_b}$$

$$= \frac{25}{I_c} h_{fe}$$

$$= h_{fe} r_e$$

$$A_v = \frac{\Delta V_{out}}{\Delta V_{in}}$$

$$= -\frac{\Delta I_c R_{out}}{\Delta I_b h_{ie}}$$

$$\boxed{A_v = -\frac{R_{out}}{h_{ie}} h_{fe}} \quad \textbf{Voltage gain}$$

$$= -\frac{R_{out}}{r_e}$$

$$= -R_{out} \frac{I_C}{25} \quad I_c \text{ in mA}$$

Sign indicates phase shift. If ΔV_{in} is positive, ΔI_b is positive and ΔI_c is positive. If ΔI_c is +ve, then I_c increases slightly, which means that V_c increases. Thus $\Delta V_{out} = \Delta V_{ce}$ must be negative.

Steps in AC analysis:

1. find h_{ie} (or r_e)

2. find R_{out}

3. calculate A_v

4. calculate low frequency response at 3 dB point
$$1 = R_{in} \omega C$$

6.8 Bypass capacitor

The AC equivalent circuit *without* bypass capacitor

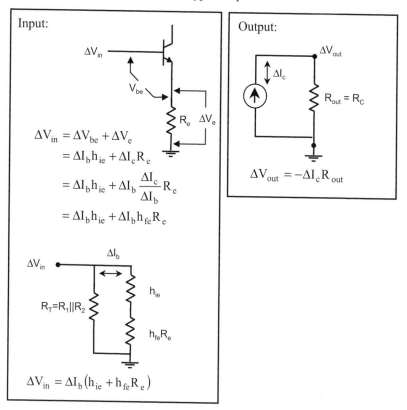

Input:

$$\Delta V_{in} = \Delta V_{be} + \Delta V_e$$
$$= \Delta I_b h_{ie} + \Delta I_c R_e$$
$$= \Delta I_b h_{ie} + \Delta I_b \frac{\Delta I_c}{\Delta I_b} R_e$$
$$= \Delta I_b h_{ie} + \Delta I_b h_{fe} R_e$$

Output:

$$\Delta V_{out} = -\Delta I_c R_{out}$$

$R_{out} = R_C$

$$\Delta V_{in} = \Delta I_b \left(h_{ie} + h_{fe} R_e \right)$$

$R_T = R_1 \| R_2$

Voltage gain *without* bypass capacitor:

$$A_v = \frac{\Delta V_{out}}{\Delta V_{in}}$$

$$= -\frac{\Delta I_c R_{out}}{\Delta I_b \left(h_{ie} + h_{fe} R_e \right)}$$

$$= -\frac{h_{fe} R_{out}}{h_{ie} + h_{fe} R_e}$$

Voltage gain *with* bypass capacitor:

$$A_v = -\frac{h_{fe} R_{out}}{h_{ie}}$$

compare

6.9 Amplifier design

Designing an amplifier involves choosing components to give the required gain A_v and input resistance R_{in}. Also, one must specify the low frequency cut-off so as to select appropriate values of coupling capacitors. The general circuit, with a fully stabilised bias, is:

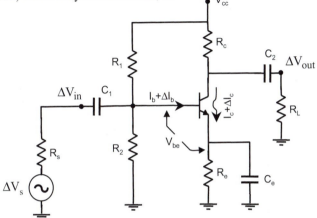

The following procedure allows the values of various components to be established in a systematic way.

1. Determine $V_{cc} = A_v/20$

2. Determine I_c from h_{ie} and h_{fe}

3. Determine I_b from $I_c/I_b = h_{fe}$

4. Determine R_c

5. Determine $V_{ce} = V_{cc}/2$

6. Determine R_e from V_e and I_c

7. Determine V_{bb} from V_{be} and V_e

8. Determine R_1 and R_2

9. Determine C_e

10. Determine C_1 and C_2

6.10 Amplifier design - input resistance

Now, $R_{in} = R_T \| h_{ie}$ If R_T is made much larger than h_{ie}, then the input resistance is dominated by the value of h_{ie}. There are other constraints on the value of R_T thus keeping R_{in} as a strong function of h_{ie} enables us to adjust h_{ie} to the desired R_{in}.

One of the most important features of the bias of an amplifier is to obtain a Q point which is at the centre of the load line (to allow maximum swing of the output V_{out}). This means that the selection of h_{ie} (for the desired R_{in}) also fixes the collector current I_c under quiescent conditions (i.e. the value of I_c when there is no input signal).

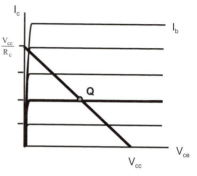

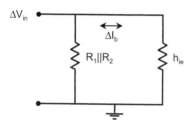

1. Thus, for a design value of h_{ie}, I_c is found from:

$$h_{ie} = \frac{h_{fe}\, 25}{\boxed{I_c}}$$

This is the first step of our step-by-step procedure for designing a common emitter amplifier.

6.11 Amplifier design – V_{cc}, I_b

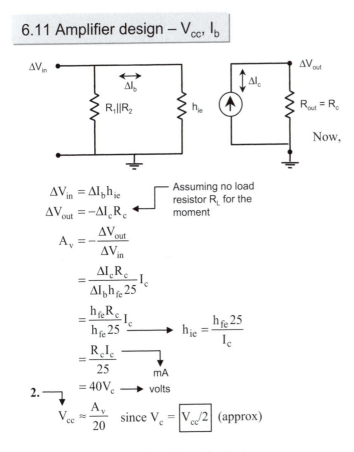

$$\Delta V_{in} = \Delta I_b h_{ie}$$
$$\Delta V_{out} = -\Delta I_c R_c$$

Assuming no load resistor R_L for the moment

$$A_v = -\frac{\Delta V_{out}}{\Delta V_{in}}$$

$$= \frac{\Delta I_c R_c}{\Delta I_b h_{fe} 25} I_c$$

$$= \frac{h_{fe} R_c}{h_{fe} 25} I_c \longrightarrow h_{ie} = \frac{h_{fe} 25}{I_c}$$

$$= \frac{R_c I_c}{25} \quad \text{mA}$$

$$= 40 V_c \longrightarrow \text{volts}$$

2.

$$V_{cc} \approx \frac{A_v}{20} \quad \text{since } V_c = \boxed{V_{cc}/2} \text{ (approx)}$$

Thus, V_{cc} is set by the choice of gain A_v

3.

$$h_{fe} = \frac{I_c}{\boxed{I_b}}$$

4. R_c may be calculated from:

$$V_c = I_c \boxed{R_c}$$

5. For a Q point to be in the centre
of the load line, then:

$$\boxed{V_{ce}} = V_{cc}/2$$

6.12 Amplifier design – V_{ce}, V_e, V_{bb}

Now, the bypass capacitor is an AC short to ground potential. Thus, $\Delta V_{out} = \Delta V_{ce}$. The slope of the AC load line is simply $-1/R_c$ (or $-1/R_c\|R_L$ if R_L is connected). But, it is only *changes* in V_{ce} that are described by the AC load line. Thus, we may draw a series of parallel lines, all with slope $-1/R_c$, to represent all the possible AC load lines of a circuit.

The actual AC load line appropriate for a particular circuit depends on the choice of V_{ce} (the DC bias conditions). The DC bias conditions are described by the DC load line:

$$I_c = \frac{-V_{ce}}{R_c + R_e} + \frac{V_{cc}}{R_c + R_e}$$

slope of DC load line
$= 1/(R_c + R_e)$

Thus, the AC load line for a particular circuit is the one which passes through the Q point on the DC load line.

$$\Delta V_{out} = \Delta V_{ce}$$
$$= \Delta I_c R_c$$
$$\Delta I_c = \frac{\Delta V_{ce}}{R_c} \quad \text{AC load lines}$$

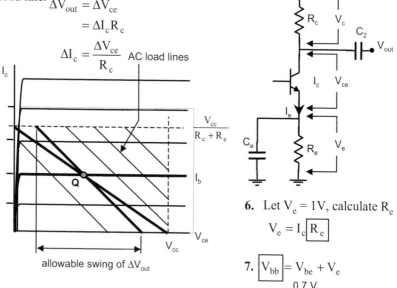

allowable swing of ΔV_{out}

6. Let $V_e = 1V$, calculate R_e
$$V_e = I_c \boxed{R_e}$$

7. $\boxed{V_{bb}} = V_{be} + V_e$
        0.7 V

Increasing R_e would reduce the y axis intercept of the DC load line and also reduce the x axis intercept of the matching AC load line. Adding a load resistor R_L would increase the slope of the AC load line while leaving the DC load line unchanged thus reducing the x axis intercept for the AC load line. Thus, to allow maximum swing on the output, R_e should be selected so as to not make V_e too large and the Q point should really be selected in the middle of the AC load line. However, selecting $V_{ce} = V_{cc}/2$ is sufficient for a first approximation (i.e. middle of DC load line).

6.13 Amplifier design – R_T

As before, the Thevenin equivalent circuit for the DC bias is:

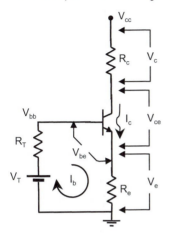

By Kirchhoff:

$$V_T = I_b R_T + \left(V_{be} + I_c R_e\right) \dots\dots\dots (1)$$

$$V_{cc} = V_c + V_{ce} + V_e \dots\dots\dots (2)$$

Now, in general:

$$h_{fe} = \frac{I_c}{I_b} \dots\dots\dots\dots\dots\dots\dots (3)$$

Thus, from (1) (2) and (3):

$$V_{ce} = V_{cc} - I_c \left(R_c + R_e\right)$$
$$= V_{cc} - h_{fe} I_b \left(R_c + R_e\right)$$
$$\Downarrow$$
$$= V_{cc} - \frac{h_{fe}\left(R_c + R_e\right)\left(V_T - V_{be}\right)}{\left(R_T + h_{fe}R_e\right)}$$

If R_T is made << than the product $h_{fe}R_e$, then V_{ce} loses its dependence on h_{fe}

$$V_{ce} = V_{cc} - \frac{\left(R_c + R_e\right)\left(V_T - V_{be}\right)}{R_e}$$

$$R_T + h_{fe}R_e \approx h_{fe}R_e$$

R_T is set so that $R_T \ll h_{fe}R_e$.
This ensures that V_{ce} is "independent" of h_{fe} and thus a stable Q point.

Open-circuit voltage:

$$V_T = V_{cc}\frac{R_2}{R_1 + R_2}$$

Equivalent resistance

$$R_T = \frac{R_1 R_2}{R_1 + R_2}$$

6.14 Amplifier design – R_1, R_2, C_1, C_2

Known: h_{ie}, h_{fe}, R_e, V_{cc}, V_c, V_{ce}, V_{bb}, I_c, I_b

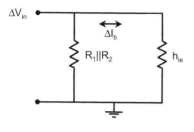

Now, in the AC circuit, R_T is in parallel with h_{ie} and thus if R_T is too small, then the input resistance of the amplifier will be reduced. This is undesirable. Thus, let $R_T \gg h_{ie}$.

For acceptable input resistance

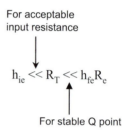

Let $\dfrac{R_T}{h_{ie}} \approx \dfrac{h_{fe}R_e}{R_T}$ This will put R_T in the middle of h_{ie} and $h_{fe}R_e$

$h_{ie} \ll R_T \ll h_{fe}R_e$

Thus $R_T \approx (h_{fe}R_e h_{ie})^{1/2}$

For stable Q point

$$R_T = \frac{R_1 R_2}{R_1 + R_2} \quad (1)$$

but

$$V_T = I_b R_T + (V_{be} + I_c R_e)$$

$$V_T = \frac{V_{cc}R_2}{R_1 + R_2} \quad (2)$$

Thus calculate V_T

8. Two eqn's, two unknowns enables R_1 and R_2 to be determined.

9. C_e from: $\dfrac{1}{\omega C_e} = 0.1 R_e$ Reactance of C_e ensures the lowest possible frequency for amplifier to operate according to design criteria.

10. C_1 and C_2 from:

$$\frac{1}{\omega C} = h_{ie}$$ Assume that $R_T = R_1 \| R_2$ is $\gg h_{ie}$ so that the input resistance is dominated by h_{ie} (which is in parallel with R_T).

Review questions

1. Analyse the DC and AC operation of the common emitter amplifier shown below, which includes the 10 kΩ load resistor.

   (a) Determine the DC bias conditions V_{bb}, I_b, I_c, V_{ce}, V_e, V_c
   (b) Determine the AC voltage gain A_v with and without the load resistor R_L present.
   (c) Determine the frequency response of the amplifier (i.e. the lowest frequency that may be amplified).

Data
   $h_{fe} = 100$
   $r_e = 25/I_c$
   $V_{be} = 0.7$ V
   $C_1 = 1$ μf

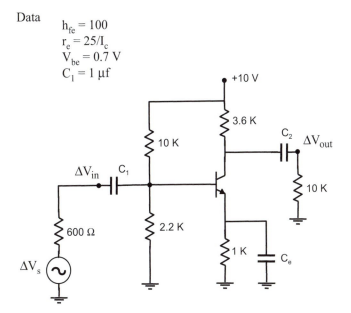

(Ans: 1.78 V, 10.8 μA, 1.08 mA, 5.04 V, 1.08 V, 3.88 V, −114.3, −155, 157 Hz)

2. A common emitter amplifier is constructed from a Darlington pair. If the current gain h_{fe} and input resistance h_{ie} of each transistor is 120 and 1.5 kΩ respectively, determine the peak-to-peak output voltage if the peak-to-peak input voltage is 10 mV. (Assume that $R_b \gg h_{ie}$ and thus the input resistance of each stage of the amplifier is h_{ie}.)

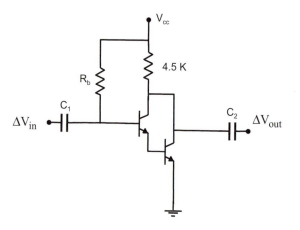

(Ans: 3.6 V)

3. The DC conditions for the common emitter amplifier shown below are: $V_{ce} = 6$ V, $I_c = 5$ mA, $V_{bb} = 2.6$ V. Determine the values of R_e, R_c and h_{fe}.

(Ans: 380 Ω, 820 Ω, 187.5)

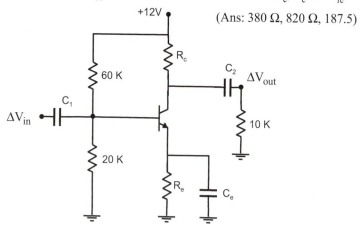

4. Consider the fully stabilised common emitter amplifier shown below. Also shown is the transistor characteristic in graphical form.

(a) Determine an equation for the load line and draw the load line on the transistor characteristic.

(b) Using the transistor characteristic, determine a value for h_{fe}.

(c) Determine the Thevenin equivalent circuit V_T, R_T of the base biasing circuit.

(d) Apply Kirchhoff's law to Thevenin equivalent circuit and obtain an expression which relates I_c to I_b and thus determine values for I_c and I_b for this circuit.

(e) Determine V_{ce} and indicate the Q point on the load line.

(f) Determine a value for r_e or h_{ie} and thus estimate the AC voltage gain A_v.

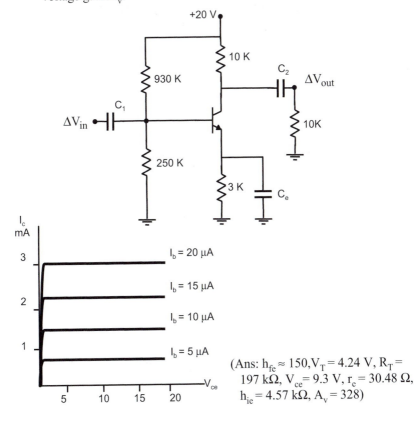

(Ans: $h_{fe} \approx 150, V_T = 4.24$ V, $R_T = 197$ kΩ, $V_{ce} = 9.3$ V, $r_e = 30.48$ Ω, $h_{ie} = 4.57$ kΩ, $A_v = 328$)

5. Design a fully stabilised common emitter amplifier with a voltage gain of 200 and a 9 V battery as the power source. The lowest frequency of operation is to be 15.92 Hz and h_{fe} for the transistor to be used is 100. Assume that an acceptable input resistance is 1 kΩ and let the voltage across R_e be set to 0.5 V and that $V_{be} = 0.7$ V.

Data

$h_{fe} = 100$
$V_{be} = 0.7$ V
$f_o = 15.9$ Hz

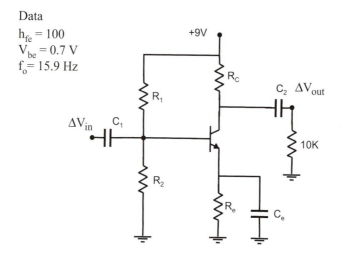

Obtain values for R_c, R_e, C_e, R_1, R_2, C_1 and C_2.

(Ans: $R_c = 2$ kΩ, $R_e = 200$ Ω, $C_e = 500$ μF, $R_1 = 30$ kΩ,
$R_2 = 5.2$ kΩ, $C_{1,2} = 10$ μF, $I_c = 2.5$ mA)

7. IO impedance

Summary

$$R_{in} \approx h_{ie}$$ Input resistance (common emitter)

$$R_{int} = \frac{\Delta V_{oc}}{\Delta I_{sc}}$$ Output resistance (common emitter)

$$A_v = \frac{h_{fe}R_e}{h_{ie} + h_{fe}R_e}$$ Common collector voltage gain

$$R_{in} \approx h_{fe}R_e$$ Input resistance (common collector)

$$R_{out} = \frac{R_s}{h_{fe}}$$ Output resistance (common collector)

7.1 CE input impedance

The **input impedance** is that seen by someone looking in at ΔV_{in}

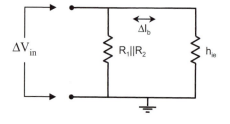

The total input impedance is $\boxed{R_{in} = R_1 \| R_2 \| h_{ie}}$

If $R_1 \| R_2$ is made larger than h_{ie} (which is usually the case), then the input resistance is approximately equal to h_{ie}

$$\boxed{R_{in} \approx h_{ie}}$$

If the signal source has some resistance R_s, then:

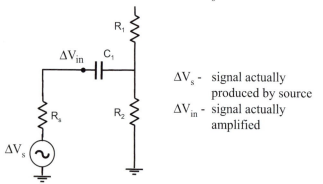

ΔV_s - signal actually produced by source

ΔV_{in} - signal actually amplified

Ideally, $R_{in} \gg R_s$ because otherwise, there is negligible ΔV_{in} at amplifier input. If R_s is large compared to R_{in}, then most of the voltage variations ΔV_s appear across R_s and not at the amplifier input.

Amplifiers should have a high **input impedance** R_{in} compared to R_s. A typical value of h_{ie} is 1 kΩ which is not all that much different to a typical R_s.

7.2 CE output impedance

Consider the output side of the CE amplifier circuit. ΔV_{out} is applied
to a load resistor R_L.

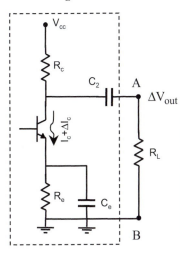

It should be possible to replace
this two-terminal "supply" by a
Norton equivalent circuit
consisting of a constant current
generator which supplies ΔI_{sc}
and an internal parallel
resistance R_{int}.

A **Norton equivalent circuit** is like
a black box that contains a special
generator that produces a variable
voltage but always produces a
constant current I equal to the
short-circuit current no matter what
value of R_L is connected to the
output terminals.

A BJT is a current generator that produces a constant ΔI_c. In the CE
amplifier circuit, this constant ΔI_c manifests itself as a voltage ΔV_{out}
across R_{int}.

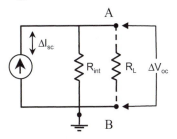

The internal resistance is the **output resistance** of the
circuit and be obtained from:

$$R_{int} = \frac{\Delta V_{oc}}{\Delta I_{sc}}$$

The output impedance is the "internal" resistance of a **Norton equivalent circuit** found from:

$$R_{out} = \frac{\Delta V_{oc}}{\Delta I_{sc}}$$

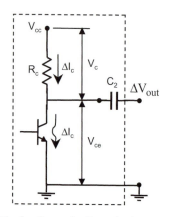

We can ignore the presence of R_e at this stage since in the AC circuit, it is shorted to ground by the bypass capacitor.

Consider the trace on an oscilloscope connected just before C_2 as there is a rise in the input voltage ($+ve\ \Delta V_{in}$).

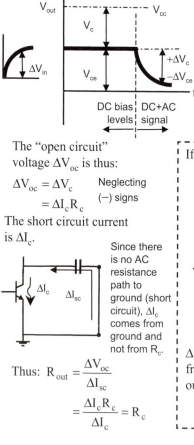

DC bias | DC+AC
levels | signal

A $+\Delta V_{in}$ (an "upswing" on the input signal) will produce a $+\Delta I_b$ and hence a $+\Delta I_C$. This results in an increase in V_c by an amount $+\Delta V_c$ across R_c and a corresponding decrease in V_{ce} by $-\Delta V_{ce}$ (bypass capacitor shorts AC signal to ground at emitter so $\Delta V_e = 0$).

$$+\Delta V_c = -\Delta V_{ce} = -\Delta V_{out}$$

The "open circuit" voltage ΔV_{oc} is thus:

$$\Delta V_{oc} = \Delta V_c \qquad \text{Neglecting}$$
$$= \Delta I_c R_c \qquad \text{(−) signs}$$

The short circuit current is ΔI_c.

Since there is no AC resistance path to ground (short circuit), ΔI_c comes from ground and not from R_c.

Thus: $R_{out} = \dfrac{\Delta V_{oc}}{\Delta I_{sc}}$

$$= \frac{\Delta I_c R_c}{\Delta I_c} = R_c$$

If a load resistor R_L connected:

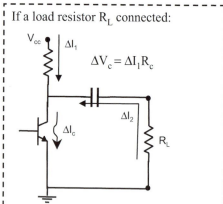

$$\Delta V_c = \Delta I_1 R_c$$

ΔI_c comes from both R_c and R_L, hence, from an AC point of view, the total output resistance is:

$$\boxed{R_{out} = R_c \parallel R_L}$$

7.3 AC Voltage gain

For a **common emitter** circuit, the AC voltage gain is:

$$A_v = -\frac{h_{fe}R_{out}}{h_{ie}}$$

For the amplifier on its own (with no load R_L connected)

$$R_{out} = R_c$$

If a load resistor is connected, then this appears in parallel with R_c thus reducing the total effective output impedance R_{out} and thus reducing the gain.

$$R_{out} = R_c \parallel R_L$$

It is desirable that the connection of a load to the amplifier does not significantly reduce the gain (since if it did, we would have to specify the load too precisely for a given circuit design). Thus, for maximum "versatility" of the circuit, we need to have R_c to be much less than any envisaged load R_L. That is, the **open circuit output resistance**, $R_{out} = R_c$, should be as low as possible.

The ideal conditions for an amplifier are thus:

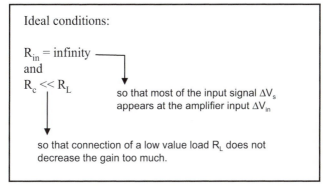

Ideal conditions:

R_{in} = infinity
and
$R_c \ll R_L$

so that most of the input signal ΔV_s appears at the amplifier input ΔV_{in}

so that connection of a low value load R_L does not decrease the gain too much.

7.4 Impedance matching

One method of increasing the gain for a particular application is to run two amplifiers together:

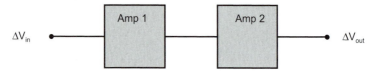

The overall gain would then be expected to be the product of the individual gains ⟶ but this is not observed.

Consider the circuit in detail:

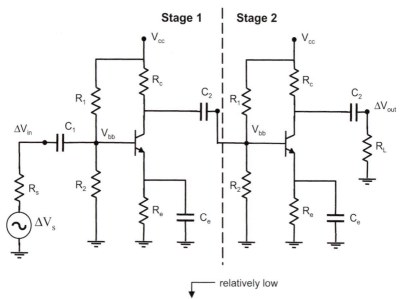

The input impedance of the 2nd stage $(R_1\|R_2\|h_{ie})$ acts as a load resistance R_L for the output of the 1st stage. This reduces the gain A_v of the first stage since $R_1\|R_2\|h_{ie}$ of 2nd stage is usually much less than R_c of first stage.

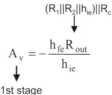

7.5 Common collector

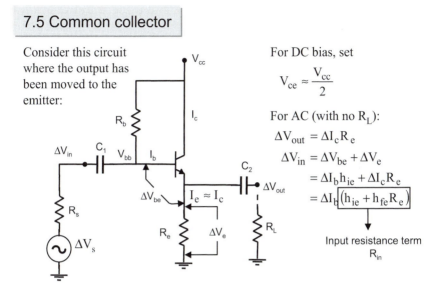

Consider this circuit where the output has been moved to the emitter:

For DC bias, set

$$V_{ce} \approx \frac{V_{cc}}{2}$$

For AC (with no R_L):

$$\Delta V_{out} = \Delta I_c R_e$$

$$\Delta V_{in} = \Delta V_{be} + \Delta V_e$$

$$= \Delta I_b h_{ie} + \Delta I_c R_e$$

$$= \Delta I_b \left(h_{ie} + h_{fe} R_e \right)$$

Input resistance term
R_{in}

Notes:

- in calculating ΔV_{in}, we have ignored R_b which is a path to (virtual) earth. However, if R_b is large compared to $h_{ie} + h_{fe}R_e$ then it may be ignored. If included, then R_b appears in parallel with $(h_{ie} + h_{fe}R_e)$

$$\Delta V_{in} = \Delta I_b \left[\left(h_{ie} + h_{fe} R_e \right) \| R_b \right]$$

- if an external load R_L is connected, then this appears in parallel with R_e and must be included with R_e in calculations.

$$\Delta V_{in} = \Delta I_b \left(h_{ie} + h_{fe} \left(R_e \| R_L \right) \right)$$

Open circuit AC voltage gain:

$$A_v = \frac{\Delta V_{out}}{\Delta V_{in}} = \frac{\Delta I_c R_e}{\Delta I_b \left(h_{ie} + h_{fe} R_e \right)}$$

$$A_v = \frac{h_{fe} R_e}{h_{ie} + h_{fe} R_e}$$ which is always less than 1

Since the output (emitter) voltage is in phase with the input, the **common collector** circuit is often called an **emitter follower**.

7.6 CC output impedance

As with a CE circuit, the CC circuit can be replaced with a Norton equivalent circuit consisting of a constant current source and a parallel resistor which is the output resistance of the circuit.

$$R_{out} = \frac{V_{opencircuit}}{I_{shortcircuit}}$$

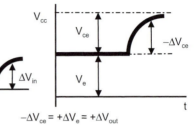

Now, since the input resistance of the common collector circuit is high, we can say that $\Delta V_s = \Delta V_{in}$, i.e. ignore the drop across R_s. Thus the open circuit

$$-\Delta V_{ce} = +\Delta V_e = +\Delta V_{out}$$

voltage is: $\Delta V_s = \Delta V_{in} = \Delta V_{out}$ since $A_v \approx 1$ and R_{in} is high.

In this circuit, the output is taken from the emitter. Hence a short circuit at the output increases the base current ΔI_b since R_e is now bypassed (this is different to the CE circuit where a short circuit at the output does not affect ΔI_b). Hence, w.r.t. ΔV_s, we have:

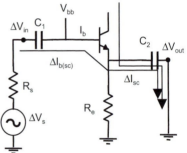

$$\Delta V_s = \Delta I_{b(sc)}(R_s + h_{ie})$$

$$= \frac{\Delta I_{sc}}{h_{fe}}(R_s + h_{ie})$$

$$\Delta I_{sc} = \frac{\Delta V_s h_{fe}}{(R_s + h_{ie})}$$

Output resistance: $R_{out} = \Delta V_{out}/\Delta I_{sc}$

$$R_{out} = \Delta V_s \frac{(R_s + h_{ie})}{\Delta V_s h_{fe}}$$

$$\boxed{R_{out} = \frac{R_s + h_{ie}}{h_{fe}}}$$

or

$$\boxed{R_{out} = \frac{R_s}{h_{fe}}}$$

If $h_{ie} \ll R_s$.

The **common collector** circuit has effectively reduced R_s by a factor of h_{fe}. The signal source with resistance R_s has been transformed into a signal source with a very low impedance R_s/h_{fe}. If V_s and R_s are actually V_{out} and R_{out} of a CE amplifier, then V_{out} can now appear to come from a low impedance output.

7.7 CC input/output impedance

Input impedance

$$\Delta V_{in} = \Delta I_b \left(h_{ie} + h_{fe} R_e \right)$$

$$\boxed{R_{in} \approx h_{fe} R_e}$$ Compare with CE amp R_{in}

May also need to
include R_b

Output impedance

$$\boxed{R_{out} = \frac{R_s}{h_{fe}}}$$

or

$$\boxed{R_{out} = \frac{R_s + h_{ie}}{h_{fe}}}$$ If h_{ie}
included

CC circuit provides a suitably low source impedance from a high
impedance source to a CE circuit.

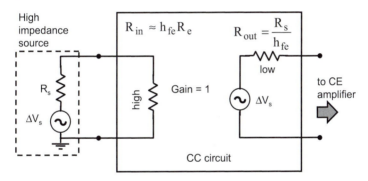

or, CC circuit provides a suitably high output impedance to CE circuit for
a low impedance load.

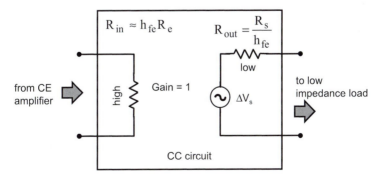

7.8 Darlington pair

The impedance matching effect depends on h_{fe}. This may be enhanced by use of **Darlington pair**:

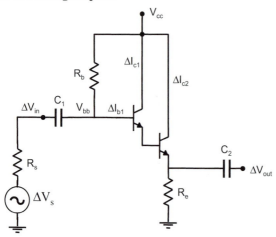

Overall current gain $h_{fe} = \Delta I_{c2}/\Delta I_{b1}$
$$= h_{fe1}h_{fe2}$$

Now, for CC:

$$R_{out} \approx \frac{R_s}{h_{fe}}$$

$$R_{in} \approx h_{fe}R_e$$

→ desirable

Thus, when h_{fe} is larger, R_{in} is larger and R_{out} is smaller.
Overall h_{fe} is increased with Darlington pair.

Review questions

1. A common collector circuit is shown below. Determine an expressions for the AC voltage gain, input resistance, and output resistance and state any desirable properties of this circuit.

Let $V_{ce} = V_{cc}/2$ for the
DC bias condition.

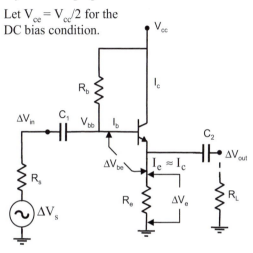

2. Calculate the input and output impedance of the common-collector circuit shown below when a source voltage of output impedance $R_s = 10$ kΩ is connected to the input. $h_{ie} = 2$ kΩ, $h_{fe} = 180$.

(Hint: in this circuit, the value of R_b is significant.)

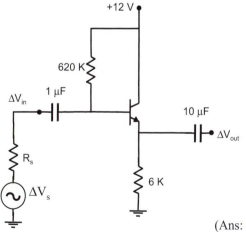

(Ans: $R_{in} = 394$ kΩ, $R_{out} = 67$ Ω)

3. Determine the component values required to bias the fully stabilised common collector amplifier shown below. Assume the emitter is held at $V_{cc}/2$ when the quiescent collector current is 1 mA and that the lowest frequency of operation is 159.2 Hz. What will be the input impedance of this circuit?

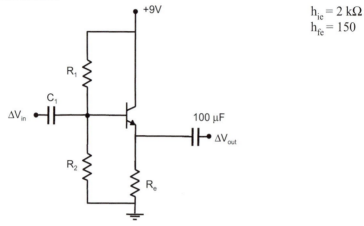

$h_{ie} = 2$ kΩ
$h_{fe} = 150$

(Ans: $R_1 = 60$ kΩ, $R_2 = 88.5$ kΩ, $R_e = 4.5$ kΩ, $I_b = 6.67$ μA)

4. The circuit below is a fully stabilised common emitter amplifier with an open circuit gain $A_v = 100$. What will be the overall gain if a pair of these amplifiers is cascaded to produce a two-stage amplifier?

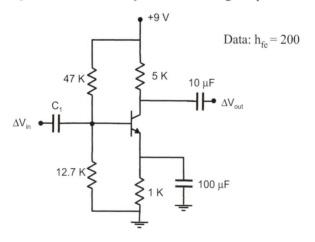

Data: $h_{fe} = 200$

(Ans: 5000)

8. Field effect transistor

Summary

$$g_m = -\frac{2\sqrt{I_{dss}}}{V_{gsoff}}\sqrt{I_d}$$ Transconductance

$$g_d = \frac{dI_d}{dV_{ds}}$$ Drain conductance

$$V_{dd} = I_d R_d + V_{ds}$$ Load line

$$I_d = -\frac{V_{ds}}{R_d} + \frac{V_{dd}}{R_d}$$

$$A_v = -g_m R_d \parallel R_L$$ Common source amplifier

8.1 Field effect transistor

pnp and npn transistors considered so far are **bi-polar junction transistors**. Bi-polar junction transistors are current controlled devices. A small I_b controls a large I_c.

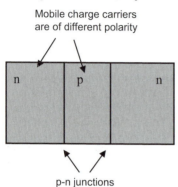

Mobile charge carriers are of different polarity

p-n junctions

Transistor action relies on the movement of two types of charge carriers hence the term "bipolar".

In a **field effect transistor**, the input voltage controls the output current. The input current is extremely small < 1 pA.

There are two main classes of FET's

JFET
Junction FET

MOSFET
Metal Oxide
Semiconductor FET

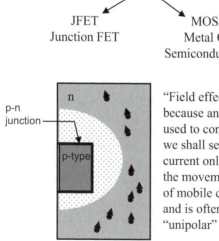

p-n
junction

n

p-type

"Field effect" so-called because an electric field is used to control current. As we shall see, the transistor current only depends on the movement of one type of mobile charge carrier and is often termed a "unipolar" device.

8.2 JFET

When the gate is made negative, ($V_{gs} < 0$) the p-n junction is in reverse bias and a depletion region develops. The p-type gate is heavily doped compared to the n-type bar, thus, most of the depletion region exists within the bar. Because the drain is +ve w.r.t. the source, the depletion layer is wider at the top than at the bottom.

i.e. the region of the p-n junction near the top of the bar near the drain is in greater reverse bias than the region at the bottom near the source.

Since the p-n junction is always in reverse bias, the current I_g through the gate circuit is very small thus presenting a high resistance R_g in this circuit.

Advantage: high input resistance

With no voltage applied to the gate, current I_d flows from drain to source. As V_{gs} is made more negative, the channel narrows as the depletion layer widens and this constriction reduces I_d

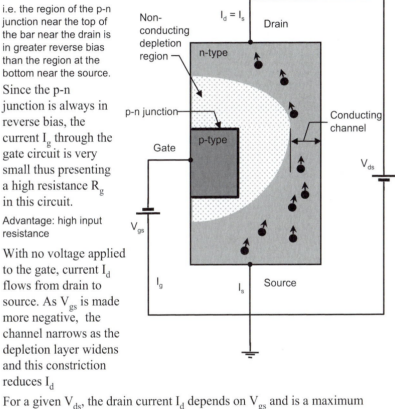

For a given V_{ds}, the drain current I_d depends on V_{gs} and is a maximum when $V_{gs} = 0$

Voltage controlled device

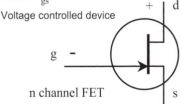

n channel FET

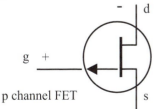

p channel FET

8.3 J-FET characteristic

How does I_d vary with V_{gs} for a fixed value of V_{ds}?

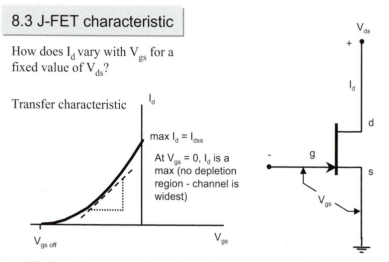

Transfer characteristic

max $I_d = I_{dss}$

At $V_{gs} = 0$, I_d is a max (no depletion region - channel is widest)

$V_{gs\ off}$

V_{gs}

V_{ds}

As V_{gs} is made more negative, channel gets narrower and I_d decreases. The decrease is given by:

$$I_d = I_{dss}\left(1 - \frac{V_{gs}}{V_{gsoff}}\right)^2$$

the slope is $\;g_m = \dfrac{dI_d}{dV_{gs}} = -\dfrac{I_{dss}}{V_{gsoff}}2\left(1 - \dfrac{V_{gs}}{V_{gsoff}}\right)$

but $\;\sqrt{\dfrac{I_d}{I_{dss}}} = \left(1 - \dfrac{V_{gs}}{V_{gsoff}}\right)$

thus $\;g_m = -\dfrac{2I_{dss}}{V_{gsoff}}\sqrt{\dfrac{I_d}{I_{dss}}}$

$$\boxed{g_m = -\frac{2\sqrt{I_{dss}}}{V_{gsoff}}\sqrt{I_d}}$$

Note, g_m comes out +ve since V_{gsoff} is −ve

Note, g_m depends on square root of drain current

transconductance
(units: mA/V = millisiemens)

Typical values:
$I_d = 5$ mA, $V_{gs} = -1$ V, $g_m = 5$ mS

Anything which makes the p-n junction more reverse-biased will affect the drain current. Thus, the transfer characteristic (and hence g_m) changes depending on the value of V_{ds} since a larger V_{ds} will result in a larger depletion region for a given V_{gs}.

8.4 JFET drain curve

Operating region: at larger V_{ds}, depletion region is wide enough to appreciably restrict current flow through channel and I_d is now a constant for a given V_{gs}

I_d mA

$V_{gs} = 0$

Ohmic region (acts like a resistor - note different slope of characteristic curves in ohmic region)

2.0

1.5

V_{gs} more negative

1.0

0.5

V_{gsoff}

I_d increases linearly with increasing V_{ds}. For small V_{ds}, not enough reverse bias w.r.t. gate for depletion layer to restrict channel.

break-down voltage

V_{ds}

V_p is also the value of V_{ds} required to bring transistor into pinch-off region with $V_{gs} = 0$

small +ve slope in operating region is called the **drain conductance**.

$V_{gsoff} = V_p$, **pinch-off voltage** which is required to block channel entirely so that $I_d = 0$

$$g_d = \frac{dI_d}{dV_{ds}}$$

important limitation in JFET circuits

Net result is I_d fairly constant

In the operating region, there are two competing conditions:

- increasing V_{ds} tends to increase I_d (just like in an ordinary resistor)
- but increasing V_{ds} makes junction more reversed-biased and thus causes narrowing of conduction channel which tends to decrease I_d

8.5 JFET load line

Consider this circuit:

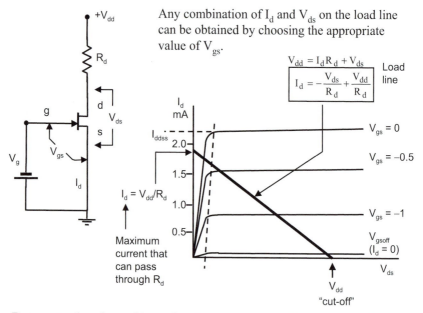

Any combination of I_d and V_{ds} on the load line can be obtained by choosing the appropriate value of V_{gs}.

$$V_{dd} = I_d R_d + V_{ds}$$
$$I_d = -\frac{V_{ds}}{R_d} + \frac{V_{dd}}{R_d}$$

Load line

$V_{gs} = 0$

$V_{gs} = -0.5$

$V_{gs} = -1$

V_{gsoff} $(I_d = 0)$

V_{ds}

$I_d = V_{dd}/R_d$

Maximum current that can pass through R_d

V_{dd}
"cut-off"

But, we need to choose V_{gs} so that Q point on the load line is within the operating region, e.g. cannot have $V_{gs} > 0$. Also, it is best to have V_{gs} in a "linear" region of the transfer characteristic to prevent distortion of large AC signals.

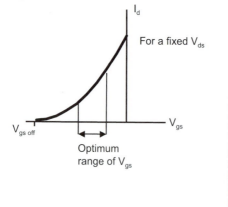

I_d

For a fixed V_{ds}

$V_{gs\,off}$

V_{gs}

Optimum range of V_{gs}

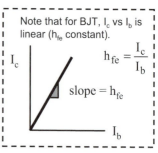

Note that for BJT, I_c vs I_b is linear (h_{fe} constant).

$$h_{fe} = \frac{I_c}{I_b}$$

slope $= h_{fe}$

I_c

I_b

8.6 JFET biasing

Now, use of a separate voltage source at
gate is not desirable, hence, consider this
circuit.

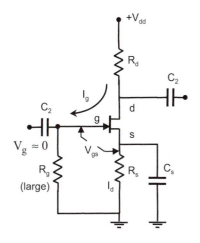

- Here, source is maintained +ve w.r.t. earth by inclusion of R_s.
- Gate is −ve w.r.t. the source by setting its potential V_g close to earth
  (i.e. it can be considered to be 0 V).
- R_g is present to offer a controlled high AC input resistance, and also
  to provide a path for the small leakage current I_g through the reverse
  bias junction to earth. Even though R_g is made large for large input
  resistance, V_g is very small because I_g is extremely small.

Circuit is self-stabilising. If I_d increases, voltage across R_s increases and
hence gate becomes more negative w.r.t. source tending to reduce I_d.

8.7 JFET amplifier biasing

DC circuit:

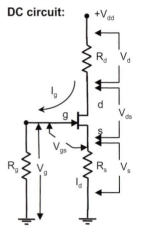

The voltages down the right hand side of the circuit are:

$$V_{dd} = V_d + V_{ds} + V_s$$

But, $V_s = I_d R_s$ since $V_g \approx 0$

$$= -V_{gs}$$

The DC value of V_{gs} is normally chosen so that the AC signal input has no possibility of producing a total voltage which makes the gate positive w.r.t. source. For an AC signal with an amplitude of 500 mV, V_{gs} should be set to about −2.0 V.

The resistor R_g provides a path for any leakage current to pass to earth and not cause any potential change at the gate which would affect the DC bias conditions. R_g should be as high as possible to maintain a high input impedance but not too high so that any leakage current produces a significant gate voltage. A value of 1 to 3 MΩ is usually sufficient.

The choice of V_{gs} also fixes the value of I_d thus the voltage across the drain resistor V_d is found from I_d and R_d:

$$V_d = I_d R_d$$

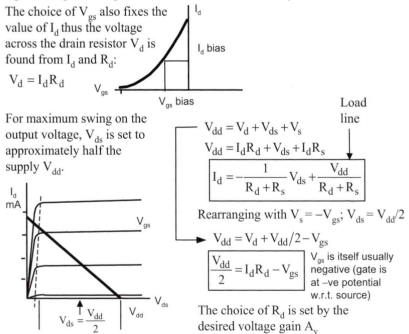

For maximum swing on the output voltage, V_{ds} is set to approximately half the supply V_{dd}.

$$V_{dd} = V_d + V_{ds} + V_s$$
$$V_{dd} = I_d R_d + V_{ds} + I_d R_s$$

$$\boxed{I_d = -\frac{1}{R_d + R_s} V_{ds} + \frac{V_{dd}}{R_d + R_s}}$$

Rearranging with $V_s = -V_{gs}$; $V_{ds} = V_{dd}/2$

$$V_{dd} = V_d + V_{dd}/2 - V_{gs}$$

$$\boxed{\frac{V_{dd}}{2} = I_d R_d - V_{gs}}$$ V_{gs} is itself usually negative (gate is at −ve potential w.r.t. source)

The choice of R_d is set by the desired voltage gain A_v

8.8 Common source amplifier

Fully stabilised **common source** FET amplifier:

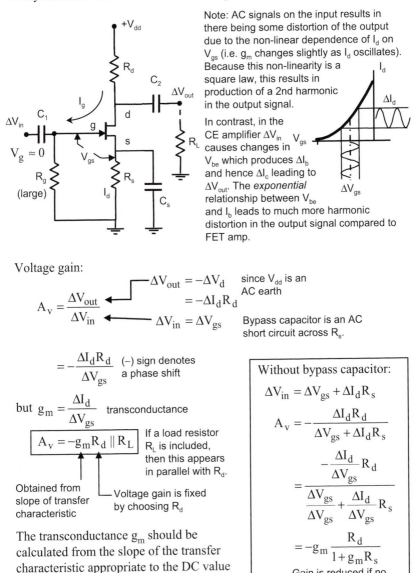

Note: AC signals on the input results in there being some distortion of the output due to the non-linear dependence of I_d on V_{gs} (i.e. g_m changes slightly as I_d oscillates). Because this non-linearity is a square law, this results in production of a 2nd harmonic in the output signal.

In contrast, in the CE amplifier ΔV_{in} causes changes in V_{be} which produces ΔI_b and hence ΔI_c leading to ΔV_{out}. The *exponential* relationship between V_{be} and I_b leads to much more harmonic distortion in the output signal compared to FET amp.

Voltage gain:

$$A_v = \frac{\Delta V_{out}}{\Delta V_{in}}$$

$$\Delta V_{out} = -\Delta V_d \quad \text{since } V_{dd} \text{ is an AC earth}$$

$$= -\Delta I_d R_d$$

$$\Delta V_{in} = \Delta V_{gs} \quad \text{Bypass capacitor is an AC short circuit across } R_s.$$

$$= -\frac{\Delta I_d R_d}{\Delta V_{gs}} \quad \text{(–) sign denotes a phase shift}$$

but $g_m = \dfrac{\Delta I_d}{\Delta V_{gs}}$ transconductance

$$\boxed{A_v = -g_m R_d \parallel R_L}$$

If a load resistor R_L is included, then this appears in parallel with R_d.

Obtained from slope of transfer characteristic

Voltage gain is fixed by choosing R_d

The transconductance g_m should be calculated from the slope of the transfer characteristic appropriate to the DC value of V_{ds} (i.e. $V_{ds} = V_{dd}/2$) and hence I_d.

Without bypass capacitor:

$$\Delta V_{in} = \Delta V_{gs} + \Delta I_d R_s$$

$$A_v = -\frac{\Delta I_d R_d}{\Delta V_{gs} + \Delta I_d R_s}$$

$$= \frac{-\dfrac{\Delta I_d}{\Delta V_{gs}} R_d}{\dfrac{\Delta V_{gs}}{\Delta V_{gs}} + \dfrac{\Delta I_d}{\Delta V_{gs}} R_s}$$

$$= -g_m \frac{R_d}{1 + g_m R_s}$$

Gain is reduced if no bypass capacitor

8.9 Drain conductance g_d

To obtain a significant voltage gain, a reasonably large value for R_d is required.

$$A_V = -g_m R_d$$

For R_d about 10 kΩ, A_V is about 30.

In the BJT, the transistor characteristic is flat, but for a JFET, the small +ve slope in the operating region *may* be significant compared to the value of R_d.

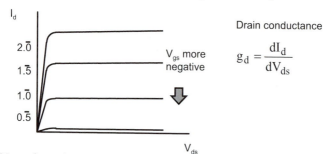

Drain conductance

$$g_d = \frac{dI_d}{dV_{ds}}$$

V_{gs} more negative

Now, from the transfer characteristic,

$$g_m = \frac{\Delta I_d}{\Delta V_{gs}}$$

$$\Delta I_d = g_m \Delta V_{gs}$$

But for a real JFET, ΔI_d depends on the change in V_{ds} as well as ΔV_{gs} because of the non-zero drain conductance g_d.

$$A_V = -g_m R_d \,\|\, \frac{1}{g_d}$$

$$= -g_m \frac{R_d \dfrac{1}{g_d}}{R_d + \dfrac{1}{g_d}}$$

$$= -g_m R_d \left(\frac{1}{1 + R_d g_d} \right)$$

If included in the analysis, then resistance of $1/g_d$ appears in parallel with R_d (and R_L) in determining the gain.

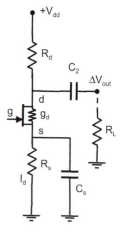

$$\boxed{A_V = -\frac{g_m R_d}{1 + R_d g_d}}$$

Note, the gain is usually expressed as a positive number, but should be used in these formulas as a negative number, eg. $A_V = -20$ for a gain of 20.

8.10 Amplifier design

Lowest operating frequency

The capacitors C_1 and C_2 serve to isolate the DC bias from the signal source and output device while allowing the AC signal to pass through.

The **3 dB point** is when $\dfrac{\Delta V_{out}}{\Delta V_{in}} = \dfrac{1}{\sqrt{2}}$

For RC high and low pass filters, this occurs when $R\omega C = 1$

The 3 dB point fixes the lowest allowable frequency of operation. If frequency of input (or output) signal is less than this condition, then too much of the signal is attenuated by the circuit (i.e. too much AC signal is blocked by the capacitor and thus becomes unavailable for amplification).

At the 3 dB point, we have, for the lowest frequency of operation ω:

C_g and C_o calculated from:
$$1 = \omega C_g R_g$$

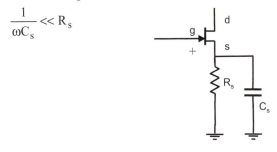

Bypass capacitor

The bypass capacitor C_s allows the circuit to be self-stabilising for DC bias fluctuations while not offering any resistance to earth for AC signals. Thus, the capacitive reactance should be made much smaller than R_s.

$$\frac{1}{\omega C_s} \ll R_s$$

8.11 Amplifier design procedure

Given or known parameters: • voltage gain A_v
                                          • transfer characteristic

1. Estimate values of V_{gs} and I_d for a point in the centre of
    the "linear" region of the characteristic and determine
    I_{dss} and V_{gsoff}

    I_d bias

2. Determine the transconductance g_m

    $$g_m = \frac{2\sqrt{I_{dss}}}{V_{gsoff}}\sqrt{I_d}$$

    V_{gs} bias

3. Determine the drain conductance g_d by measuring the
    slope of the operating region on the transistor
    characteristic at the chosen values of V_{gs} and I_d

4. Calculate R_d from: $A_v = -g_m R_d$ or $A_v = -\dfrac{g_m R_d}{1 + g_d R_d}$

5. Calculate R_s from: $R_s = \dfrac{-V_{gs}}{I_d}$

6. Let $V_{ds} = V_{dd}/2$ and thus calculate V_{dd} from: $\dfrac{V_{dd}}{2} = I_d R_d - V_{gs}$

7. Choose C_1 and C_2 so that $R_g \omega C = 1$ and
    C_s so that $1/\omega C_s \ll R_s$

    Note: V_{gs} is negative,
    and must be inserted
    into this formula as is,
    for example:

    $$\frac{V_{dd}}{2} = I_d R_d - (-1.5)$$

In JFET, I_d is normally made smaller
than I_c in a BJT circuit so as to avoid
the use of a large voltage supply V_{dd}.
For example, if I_d is 2 mA, and $R_s + R_d$
= 20 kΩ, then V_{dd} = 80 V.

↓

Need to lower this to some
reasonable value (e.g. 20 V)
by sacrificing gain or lowering
I_d (more negative V_{gs}) and
accepting more distortion.

8.12 MOSFETS

Metal Oxide Semiconductor, Field Effect Transistor. A **MOSFET** is
similar to a **JFET** but has the gate insulated from the channel.

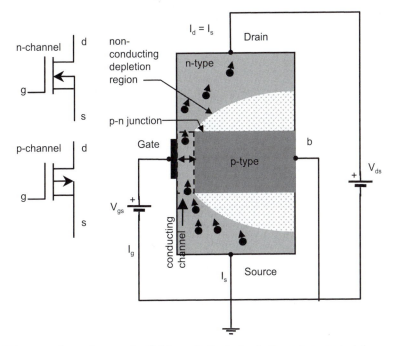

When gate is made +ve, the field repels the holes in the p-type material
leaving behind what is essentially n-type silicon which permits charge
carriers to move from source to drain (conventional current from drain to
source).

Or, attracts electrons from n-type to fill holes in p-type (similar to BJT).
When all holes are filled, conducting channel is formed.

When $V_{gs} = 0$, $I_d = 0$. MOSFET is normally OFF when $V_{gs} = 0$ (unlike JFET).
The drain current depends on gate voltage. Like JFET, drain current
increases when the gate voltage is made more positive.

Very high input impedance, 10^{11} Ω

8.13 CMOS

Complementary Metal Oxide Semiconductor
A **CMOS** chip contains both p- and n-channel **MOSFET**s on the same chip.

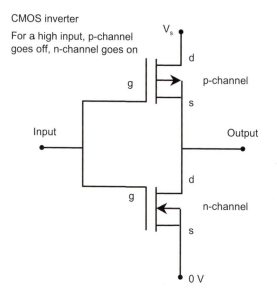

CMOS inverter
For a high input, p-channel
goes off, n-channel goes on

Characteristics of CMOS digital circuits:
- very compact allowing high packing density on a single chip
- low power consumption
- accept large range of power supply voltages (+3 - +15 V)
- slow speed compared to TTL

CMOS chips are used in computer circuitry because they only consume power when they change state. When they are in a particular state, no current flows and no power is consumed. This makes them ideal for portable electronic equipment.

Very high input impedance means very little current draw from inputs.

Review questions

1. Explain, with the aid of a diagram, as briefly as possible, the principle of operation of a JFET. In your explanation, describe the effect of various applied voltages across p-n junctions how transistor action is obtained.

2. In the circuit below, V_{ds} is measured by a student at 29 V. Calculate the DC bias conditions (V_{gs}, I_d) given that $V_{dd} = 40$ V, $R_s = 300$ Ω, $R_g = 2$ MΩ, and $R_D = 6$ kΩ. Also calculate the AC voltage gain and the peak-to-peak output voltage when $\Delta V_{in} = 30$ mV.

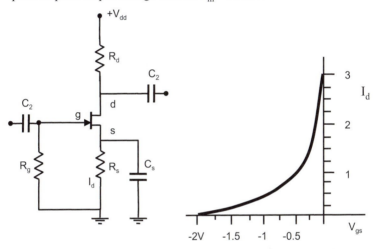

(Ans. $I_d = 1.74$ mA, $V_{gs} = -0.5$ V, $g_m = 3.48$ mS, $A_v = 20.9$, $\Delta V_{out} = 624$ mV)

3. An FET has an input resistance of 1000 MΩ and a transconductance of 4 mS. The drain conductance is 100 μS at the operating point where $V_{ds} = 14$ V, $I_d = 2$ mA, $V_{gs} = -2$ V. Draw a circuit for a single stage amplifier and determine the values of as many components as possible if a 30 V supply is available. Compare the voltage gain with and without the drain conductance being taken into consideration.

(Ans: $R_d = 7$ kΩ, $R_s = 1$ kΩ, $A_v = 28$, $A_v = 16.5$)

4. An FET is used in an amplifier circuit with a load resistor of $R_d = 40$ kΩ. The gain is measured at 40. Calculate the output resistance and the transconductance of the transistor if the gain drops to 30 when the load resistance is halved.

(Ans: 20 kΩ, 3 mS)

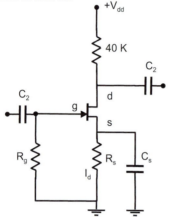

5. An FET amplifier is used as the first stage of a two stage amplifier as shown below. The second stage is a simple-biased common emitter amplifier. Calculate the overall gain of the complete amplifier circuit.

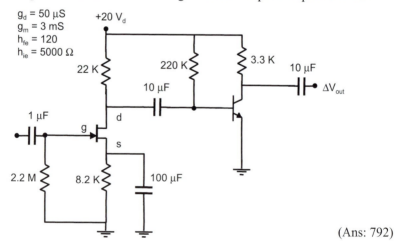

(Ans: 792)

9. High frequency

Summary

$$\frac{1}{r_e \omega_T C} \approx 1 \qquad \text{Transition frequency}$$

$$\frac{1}{Z_{in}} = \frac{1}{h_{ie}} + j\omega(C_{be} + AC_{cb}) \qquad \text{Miller effect}$$

9.1 BJT capacitance

In a transistor, reverse-biased c-b junction behaves like a capacitor since the depletion region is an insulator. The capacitance is given as C_{cb}.

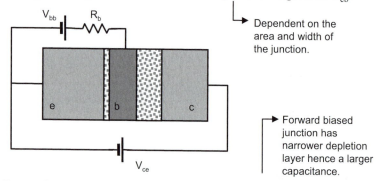

↳ Dependent on the area and width of the junction.

Forward biased junction has narrower depletion layer hence a larger capacitance.

Forward bias b-e junction also possesses capacitance C_{be}
- some of this capacitance is due to the presence of the depletion region
- some of this capacitance is due to the finite speed of diffusion of charge carriers across the junction and is significant when the input signal changes rapidly.

Both these capacitances restrict the **high frequency** response of the transistor. An equivalent circuit for a transistor is thus:

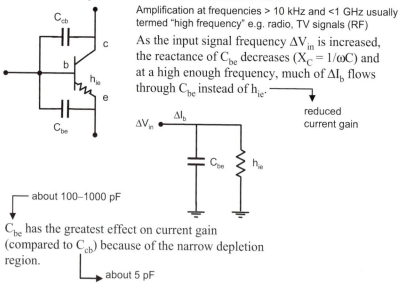

Amplification at frequencies > 10 kHz and <1 GHz usually termed "high frequency" e.g. radio, TV signals (RF)

As the input signal frequency ΔV_{in} is increased, the reactance of C_{be} decreases ($X_C = 1/\omega C$) and at a high enough frequency, much of ΔI_b flows through C_{be} instead of h_{ie}. ──┐
                                                                                          ↓
                                                                              reduced current gain

⌐ about 100–1000 pF
↓

C_{be} has the greatest effect on current gain (compared to C_{cb}) because of the narrow depletion region.

↳ about 5 pF

9.2 Transition frequency

At high frequencies, the base current I_b is diverted more and more into C_{be} (since X_c decreases) and the effective current gain of the transistor is reduced.

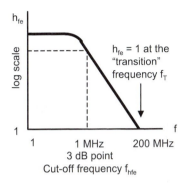

$h_{fe} = 1$ at the "transition" frequency f_T

1 MHz   200 MHz
3 dB point
Cut-off frequency f_{hfe}

The transition frequency marks the point where the transistor cannot be used as an amplifier. It depends upon the collector current since the input resistance h_{ie} is a function of the collector current.

Thus, an increase in I_c results in a decrease in h_{ie} and since C_{be} is in parallel with h_{ie}, there is an increase in the transition frequency. At higher collector currents, the capacitance C_{be} begins to rise and dominate any effect of decreasing h_{ie} causing transition frequency to fall with increasing I_c

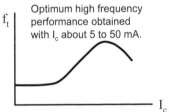

Optimum high frequency performance obtained with I_c about 5 to 50 mA.

C_{be} rises due to a change in the apparent size of the base region as the depletion layer is shifted off-centre towards the emitter. This effects the time taken for charge carriers to pass through the base which at high frequencies, is an effective capacitance.

Let h_{fe} be the current gain of the transistor without any frequency effects. As the frequency gets higher, the overall current gain approaches 1 meaning that the current through the base which actually gets amplified is reduced to $\Delta I_b / h_{fe}$. The remainder of ΔI_b goes through C_{be}.

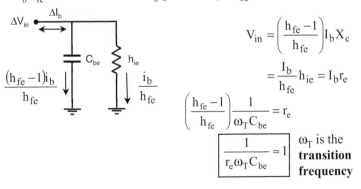

$$V_{in} = \left(\frac{h_{fe} - 1}{h_{fe}} \right) I_b X_c$$

$$= \frac{I_b}{h_{fe}} h_{ie} = I_b r_e$$

$$\left(\frac{h_{fe} - 1}{h_{fe}} \right) \frac{1}{\omega_T C_{be}} = r_e$$

$$\frac{1}{r_e \omega_T C_{be}} \approx 1$$

ω_T is the **transition frequency**

9.3 C_{cb} Common emitter amplifier

In a CE voltage amplifier, the output voltage ΔV_{out} is essentially the collector voltage. The output signal, at high frequencies, can thus be passed back to the input side of the transistor through C_{cb}.

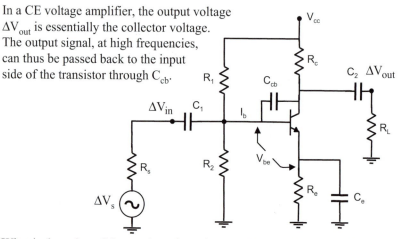

What is the value of C_{cb} as viewed purely from the amplifier input ΔV_{in}?

$$\Delta V_{cb} = \Delta V_{in} - \Delta V_{out}$$

but $\Delta V_{out} = -A_v \Delta V_{in}$

thus $\Delta V_{cb} = \Delta V_{in}(A_v + 1)$

now $C_{cb} = \dfrac{q}{V_{cb}}$

therefore $\dfrac{q}{C_{cb}} = \Delta V_{in}(A + 1)$

$$q = \Delta V_{in}(A_v + 1)C_{cb}$$

the capacitance as seen from the input ΔV_{in}, w.r.t. earth, is:

$$C_{eff} = \dfrac{q}{\Delta V_{in}}$$

$$= \dfrac{\Delta V_{in}(A_v + 1)C_{cb}}{\Delta V_{in}}$$

$$= (A_v + 1)C_{cb}$$

$$\approx A_v C_{cb} \quad \text{much larger than first expected.}$$

From the point of view of ΔV_{in}, the capacitance C_{cb} actually appears to be much larger due to feedback effect from output side of circuit. This is the **Miller effect** and is the most significant barrier to high frequency voltage amplification.

9.4 Miller effect

Consider this equivalent circuit of a CE amplifier

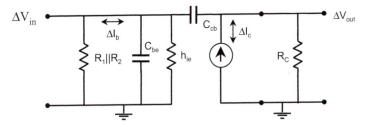

Equivalent circuit can be shown with C_{cb} replaced by effective capacitance AC_{cb}:

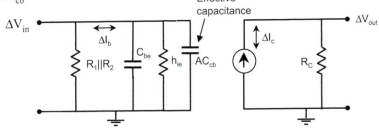

The internal capacitances C_{be} and AC_{cb} serve to decrease the input impedance of the amplifier circuit. The total input impedance becomes frequency dependent and is given by:

$$\frac{1}{Z_{in}} = \frac{1}{h_{ie}} + j\omega\left(C_{be} + AC_{cb}\right)$$
(neglecting $R_1\|R_2$)

Thus, at high frequencies, the measured voltage gain of a CE amplifier circuit will decrease as the input impedance of the circuit becomes smaller in comparison to the output impedance of the source (i.e. less of ΔV_{in} appears across the input to the amplifier and more across the internal resistance of the source).

What affects the value of C_{cb}?

depends on charge
carrier mobility

• the dimensions of the depletion layer
• the speed of diffusion of charge carriers across the junction.

How can these effects be minimised?

9.5 Output capacitance

At high frequencies, capacitances of various items connected to the output serve to short the output signal to earth thus causing an apparent decrease in voltage gain:

- input capacitance of load
- capacitance of connecting wires
- etc

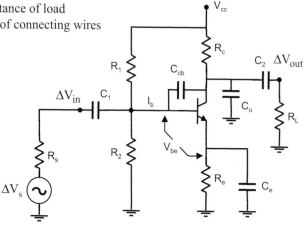

If the frequency of the signal to be amplified is increased, and ΔV_{out} decreases without any decrease in ΔV_{in}, then the chances are that a stray capacitance C_o is shunting the signal to earth instead of into the desired output device.

- such as would happen if C_{be} and AC_{cb} were significant

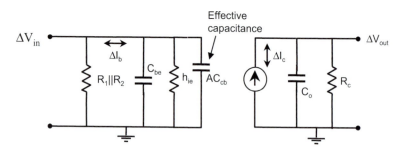

9.6 Amplification at high frequencies

One example of a circuit with BJT transistor for high frequency applications is the **common base** amplifier circuit.

How does it work?

Input signal feeds into the resistance of the base-emitter junction as seen from the emitter. i.e. r_e which is low.

$$r_e = \frac{25}{I_e} \rightarrow \text{DC emitter current in mA and } I_c = I_e$$

Thus, the circuit has a low input resistance (not desirable). However, the input signal only sees capacitance C_{be} and not $A_v C_{cb}$ thus Miller effect does not affect ΔV_{in}.

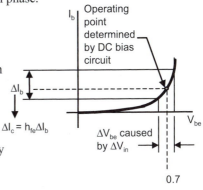

This amplifier works somewhat backwards in that the input signal causes ΔV_{be} to oscillate from the emitter side rather than from the base side but here, when the signal ΔV_{in} increases, I_b decreases and so does I_c and thus ΔV_{out} goes up - input and output are in phase.

- Voltage gain: same as CE amplifier
- Low input impedance (not so good but...)
- useful for preamplifier for television signals from coaxial cable (low impedance source 70 Ω)
- Miller effect on gain is reduced - very desirable
- Useful high frequency response only limited by transition frequency f_T

Review questions

1. The transistion frequency for a certain transistor is specified at 320 MHz at $I_c = 12$ mA. Determine the capacitance of the base-emitter junction.

   (Ans: 239 pF)

2. Calculate the effective capacitance on the input of an amplifier $A_v = 120$ which uses a transistor with $C_{cb} = 6$ pF.

   (Ans: 726 pF)

3. What capacitance has the most effect on the gain of:

   (a) a BJT transistor;
   (b) a common emitter amplifier.

4. An amplifier circuit has the following parameters. Calculate the input impedance Z_{in} at $\omega = 200$ MHz, $h_{ie} = 500\ \Omega$, $C_{be} = 150$ pF, $C_{cb} = 5$ pF, $A_v = 200$

   (Ans: 0.59 Ω)

10. Transients

Summary

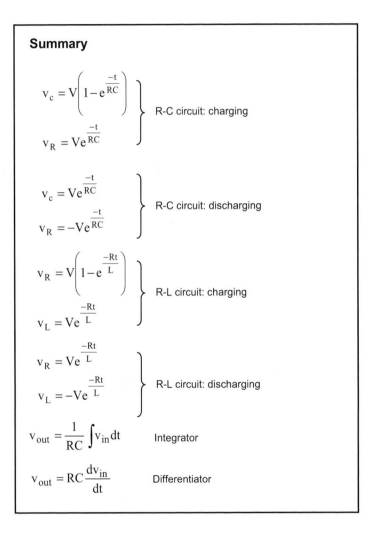

$$v_c = V\left(1 - e^{\frac{-t}{RC}}\right)$$

$$v_R = Ve^{\frac{-t}{RC}}$$

R-C circuit: charging

$$v_c = Ve^{\frac{-t}{RC}}$$

$$v_R = -Ve^{\frac{-t}{RC}}$$

R-C circuit: discharging

$$v_R = V\left(1 - e^{\frac{-Rt}{L}}\right)$$

$$v_L = Ve^{\frac{-Rt}{L}}$$

R-L circuit: charging

$$v_R = Ve^{\frac{-Rt}{L}}$$

$$v_L = -Ve^{\frac{-Rt}{L}}$$

R-L circuit: discharging

$$v_{out} = \frac{1}{RC}\int v_{in}\,dt$$

Integrator

$$v_{out} = RC\frac{dv_{in}}{dt}$$

Differentiator

10.1 R-C circuit analysis

Consider a series circuit containing a resistor R and capacitor C. If connection (a) is made, the initial potential difference across the capacitor is zero, the entire battery voltage appears across the resistor. At this instant, the current in the circuit is:

$$I_o = \frac{V}{R}$$

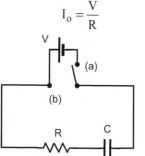

As the capacitor charges, the voltage across it increases and the voltage across the resistor correspondingly decreases (and so does the current in the circuit). After some time, the capacitor is fully charged and the entire battery voltage appears across the capacitor. The current thus drops to zero and there is no potential drop across the resistor (since there is no current).

When connection (b) is made, the capacitor discharges through the resistor. The current is initially high and then over time drops to zero.

$$v_R = iR; \quad v_C = \frac{q}{C}$$

$$V = v_R + v_C$$

Small letters signify instantaneous values

$$= iR + \frac{q}{C}$$

$$i = \frac{V}{R} - \frac{q}{RC} \quad \text{but} \quad i = \frac{dq}{dt}$$

$$\frac{dq}{dt} = \frac{V}{R} - \frac{q}{RC}$$

$$\frac{dq}{VC - q} = \frac{dt}{RC}$$

1st order differential equation - integrate both sides

$$-\ln(VC - q) = \frac{t}{RC} + \text{constant}$$

When t = 0, q = 0.

$$-\ln(VC - q) = \frac{t}{RC} - \ln VC$$

$$VC = Q_f$$

$$\ln(VC - q) - \ln VC = \frac{-t}{RC}$$

$$\ln \frac{VC - q}{VC} = \frac{-t}{RC}$$

$$1 - \frac{q}{VC} = e^{\frac{-t}{RC}}$$

At fully charged capacitor has charge Q_f and potential difference = V

$$q = VC\left(1 - e^{\frac{-t}{RC}}\right)$$

$$q = Q_f\left(1 - e^{\frac{-t}{RC}}\right)$$

$$\frac{dq}{dt} = \frac{V}{R} e^{\frac{-t}{RC}}$$

Both current in circuit and charge on capacitor are an exponential functions of time

$$= I_o e^{\frac{-t}{RC}}$$

$$= i$$

I_o is the initial current in the circuit

$$i = I_o e^{\frac{-t}{RC}}$$

10.2 Time constant and half-life

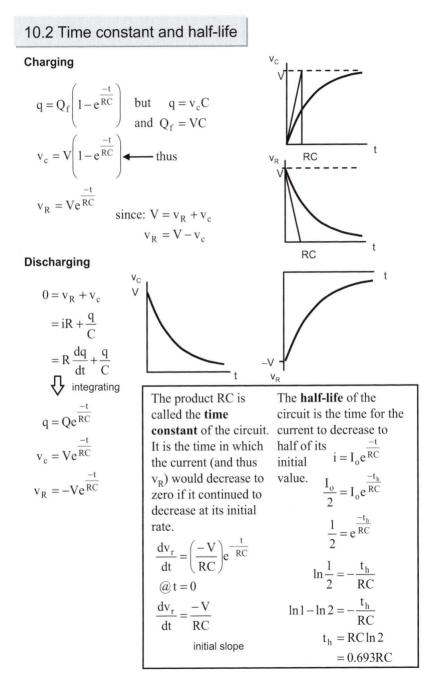

Charging

$$q = Q_f \left(1 - e^{\frac{-t}{RC}}\right) \quad \text{but} \quad q = v_c C$$
$$\text{and} \quad Q_f = VC$$

$$v_c = V\left(1 - e^{\frac{-t}{RC}}\right) \longleftarrow \text{thus}$$

$$v_R = Ve^{\frac{-t}{RC}}$$

$$\text{since:} \quad V = v_R + v_c$$
$$v_R = V - v_c$$

Discharging

$$0 = v_R + v_c$$
$$= iR + \frac{q}{C}$$
$$= R\frac{dq}{dt} + \frac{q}{C}$$
$$\Downarrow \text{integrating}$$

$$q = Qe^{\frac{-t}{RC}}$$

$$v_c = Ve^{\frac{-t}{RC}}$$

$$v_R = -Ve^{\frac{-t}{RC}}$$

The product RC is called the **time constant** of the circuit. It is the time in which the current (and thus v_R) would decrease to zero if it continued to decrease at its initial rate.

$$\frac{dv_r}{dt} = \left(\frac{-V}{RC}\right)e^{-\frac{t}{RC}}$$
$$@\, t = 0$$
$$\frac{dv_r}{dt} = \frac{-V}{RC}$$

initial slope

The **half-life** of the circuit is the time for the current to decrease to half of its initial value.

$$i = I_o e^{\frac{-t}{RC}}$$

$$\frac{I_o}{2} = I_o e^{\frac{-t_h}{RC}}$$

$$\frac{1}{2} = e^{\frac{-t_h}{RC}}$$

$$\ln\frac{1}{2} = -\frac{t_h}{RC}$$

$$\ln 1 - \ln 2 = -\frac{t_h}{RC}$$

$$t_h = RC\ln 2$$
$$= 0.693RC$$

10.3 R-C low pass filter

As capacitor charges, voltage across it (v_{out}) increases. When capacitor is fully charged, all of v_{in} appears across it and none across the resistor. As the time constant becomes smaller than the period T of the input pulse, the capacitor has more time to charge and discharge fully before the pulse changes polarity.

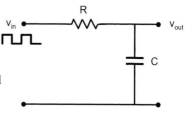

For **low pass filter**, we want RC << T to pass through low frequencies. i.e. small time constant.

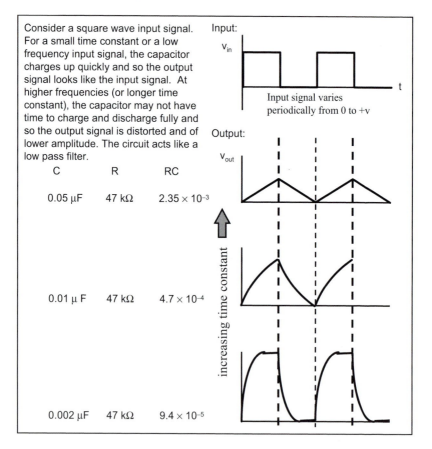

Consider a square wave input signal. For a small time constant or a low frequency input signal, the capacitor charges up quickly and so the output signal looks like the input signal. At higher frequencies (or longer time constant), the capacitor may not have time to charge and discharge fully and so the output signal is distorted and of lower amplitude. The circuit acts like a low pass filter.

Input signal varies periodically from 0 to +v

C	R	RC
0.05 μF	47 kΩ	2.35×10^{-3}
0.01 μF	47 kΩ	4.7×10^{-4}
0.002 μF	47 kΩ	9.4×10^{-5}

10.4 R-C high pass filter

As capacitor charges, voltage across it increases and voltage across resistor decreases. As time constant becomes smaller than the period T of the input pulse, the capacitor has time to charge and discharge fully and the voltage across the resistor decreases to zero.

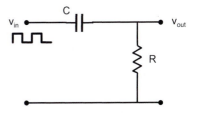

For **high pass filter**, we want RC >> T to pass through high frequencies. i.e. large time constant.

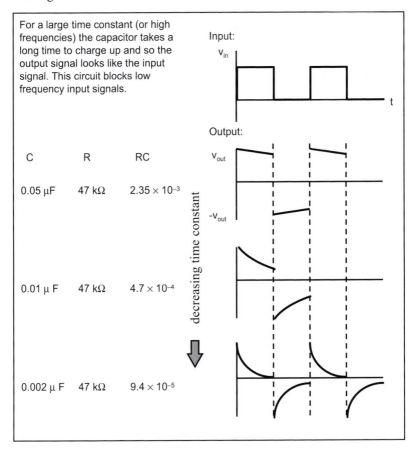

For a large time constant (or high frequencies) the capacitor takes a long time to charge up and so the output signal looks like the input signal. This circuit blocks low frequency input signals.

Input:

v_{in}

Output:

v_{out}

C	R	RC
0.05 μF	47 kΩ	2.35×10^{-3}
0.01 μF	47 kΩ	4.7×10^{-4}
0.002 μF	47 kΩ	9.4×10^{-5}

decreasing time constant

10.5 R-L circuits

Charging

$$V = v_R + v_L$$

$$V = iR + L\frac{di}{dt}$$

⇩

$$i = I\left(1 - e^{\frac{-Rt}{L}}\right)$$

$$v_R = V\left(1 - e^{\frac{-Rt}{L}}\right)$$

$$v_L = Ve^{\frac{-Rt}{L}}$$

The quantity L/R is the **time constant** of the circuit and I is the final current.

Note: as written here, we write +Ldi/dt to signify a voltage <u>drop</u> across the inductor.

$$v_L = L\frac{di}{dt}$$

$$v_R = iR$$

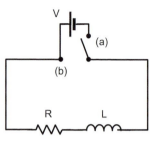

Let connection (a) be made. Because of the self induced emf, the current (and hence v_R) in the circuit does not rise to its final value at the instant the circuit is closed, but grows at a rate which depends on the **inductance** (henrys) and resistance (ohms) of the circuit.

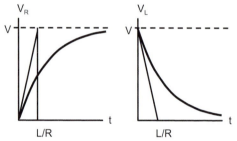

Discharging

$$0 - v_L = v_R$$

$$0 = iR + L\frac{di}{dt}$$

⇩

$$i = Ie^{\frac{-Rt}{L}}$$

$$v_R = Ve^{\frac{-Rt}{L}}$$

$$v_L = -Ve^{\frac{-Rt}{L}}$$

When connection (b) is made the current (and v_R) does not fall to zero immediately but falls at a rate which depends on L and R. The energy required to maintain the current during the decay is provided by the energy stored in the magnetic field of the conductor

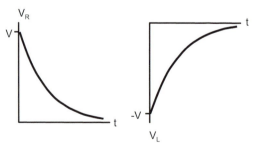

10.6 R-L filter circuits

Low pass (choke) circuit

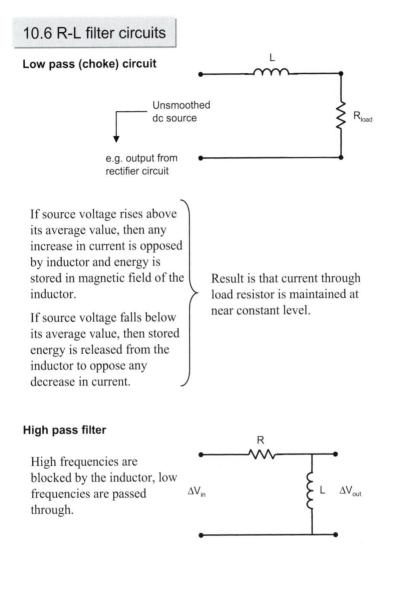

If source voltage rises above its average value, then any increase in current is opposed by inductor and energy is stored in magnetic field of the inductor.

If source voltage falls below its average value, then stored energy is released from the inductor to oppose any decrease in current.

Result is that current through load resistor is maintained at near constant level.

High pass filter

High frequencies are blocked by the inductor, low frequencies are passed through.

10.7 Transistor circuits

Transistor switch - this circuit will "square up" any repetitive waveform

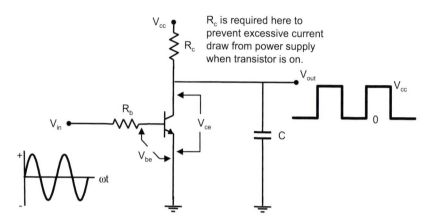

R_c is required here to prevent excessive current draw from power supply when transistor is on.

Now, consider the effect of a capacitance on the output:

If the transistor is initially off, then $V_{out} = V_{cc}$ and the capacitor will be charged also to V_{cc}

But this charging takes place through R_c, i.e. time-constant = RC, and the rate of rise of V_{out} is relatively long.

When the transistor is turned on, the V_{out} drops low and capacitor discharges through V_{ce} to earth.

Since this connection through to earth is virtually a short circuit, through the transistor from collector to emitter, then discharge time is very short.

Resulting output:

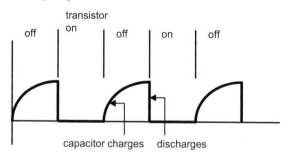

In active switching circuits, "turn-on" is faster than "turn-off".

10.8 Active pull-up

Technique to overcome slow turn-off.

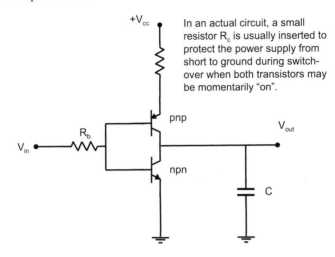

In an actual circuit, a small resistor R_c is usually inserted to protect the power supply from short to ground during switch-over when both transistors may be momentarily "on".

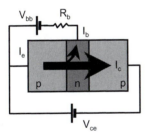

Direction of hole current is shown.

Positive input voltage:

• npn transistor base emitter junction in forward bias so transistor turns on. Capacitor can discharge to earth through collector-emitter as before.

• pnp transistor base-emitter junction in reverse bias so transistor turns off and emitter-collector becomes non-conducting. V_{cc} is thus not shorted to earth through npn transistor. Capacitor voltage is now V_{ce} for npn since pnp is off .

Zero or negative input voltage:

• pnp transistor is turned on and V_{cc} appears across capacitor and also at V_{out}. But, charging time is very short since resistance R_c may be eliminated in this circuit.

The same effect can be obtained with two npn transistors if the bases of each are driven separately and out of phase so that one switches on when the other switches off. ⟶ IC logic circuits usually employ this method since they are more easily fabricated on one semi-conductor chip.

10.9 Integrator/differentiator

Integrating circuit

$$v_{out} = \frac{q}{C}$$

$$\frac{dq}{dt} = i \therefore q = \int i\,dt$$

thus $v_{out} = \frac{1}{C} \int i\,dt$

now, $v_R = iR$

and thus $\therefore v_{out} = \frac{1}{RC} \int v_R\,dt$

Now $v_R \approx v_{in}$ when RC is large

or $v_c \ll v_r$

thus $\boxed{v_{out} = \frac{1}{RC} \int v_{in}\,dt}$ Output voltage signal is the integral of the input voltage signal.

Differentiating circuit

$$v_{out} = IR$$

$$I = \frac{dq}{dt}$$

$$= C\frac{dv_C}{dt}$$

$$v_{out} = RC\frac{dv_C}{dt}$$

For small RC capacitor charges up quickly.

$v_C \gg v_R \therefore v_C \approx v_{in}$ when RC is small

$$\boxed{v_{out} = RC\frac{dv_{in}}{dt}}$$

Output voltage is the derivative of the input voltage.

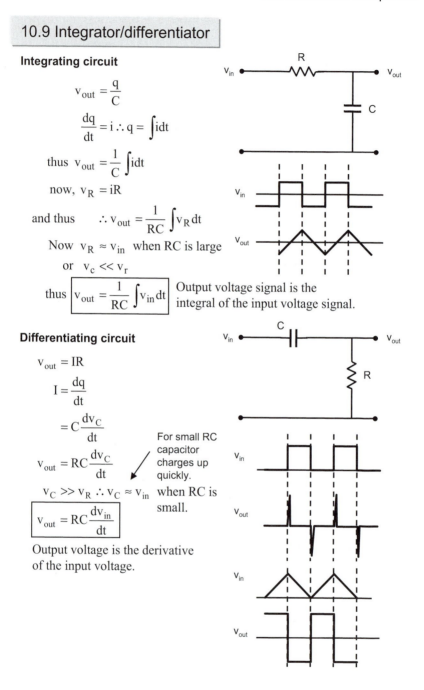

Review questions

1. A circuit containing an inductor 20 H with a resistance or 16 Ω is connected through a switch to a 24 V source. Calculate the following:
   (a) the initial current when the switch is closed;
   (b) the final steady-state current;
   (c) the initial rate of change of current when the switch is closed;
   (d) the time-constant for this circuit.

   (Ans: 0, 1.5 A, 1.2 A s^{-1}, 1.25 s)

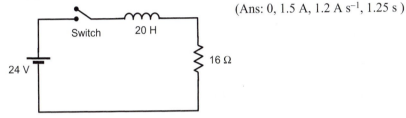

2. A capacitor of 1 μF is charged through a 50 kΩ resistor. Calculate how long it will take the capacitor to charge to 63% of the applied voltage.

   (Ans: 0.05 s)

3. Design an RC filter to give at least 3 dB attenuation for all frequencies above 10 kHz.

4. A square wave of frequency 100 kHz is applied to the following network. If the square wave amplitude is 4 V, determine the output wave from the network.

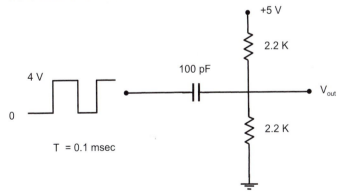

5. A conventional ignition system in a motor vehicle consists of an induction coil, the current to which is periodically switched on and off through mechanical "contact breaker points". A high voltage is induced in the secondary side of the coil when the contact breaker opens and closes.

If the resistance of the primary side of the circuit is 8 Ω and the inductance of the coil is 2.4 H, calculate the following quantities:

(a) the initial current when the contact breaker is just closed;
(b) the initial <u>rate of change</u> of current when the contact breaker is just closed;
(c) the final steady-state current;
(d) the time taken for the current to reach 95% of its maximum value.

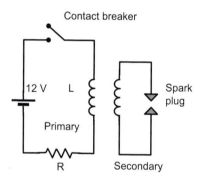

(Ans: 0, 5 A s^{-1}, 1.5 A, 0.9 s)

11. Digital electronics

Summary

Boolean algebra

$A + B = B + A$

$B \cdot A = A \cdot B$

$(A + B) + C = A + (B + C)$

$(A \cdot B) \cdot C = A \cdot (B \cdot C)$

$A + AB = A \cdot (1 + B) = A$

$A \cdot (A + B) = A$

$A \cdot (B + C) = A \cdot B + A \cdot C$

$A + (B \cdot C) = (A + B) \cdot (A + C)$

$A + A = A$

$A \cdot A = A$

$A \cdot \overline{A} = 0$

$A + \overline{A} = 1$

$\overline{\overline{A}} = A$

$0 + A = A$

$1 \cdot A = A$

$1 + A = 1$

$0 \cdot A = 0$

$A + \overline{A} \cdot B = A + B$

$A \cdot (\overline{A} + B) = A \cdot B$

De Morgan's theorem

$\overline{(A + B)} = \overline{A} \cdot \overline{B}$

$\overline{(A \cdot B)} = \overline{A} + \overline{B}$

11.1 Digital logic

Digital electronic circuits contain components which act like high speed switches that process voltage levels that are suitable for representing the binary numbers 0 and 1. These voltage levels may also represent **logic states** true and false and thus allow binary data to be processed using **Boolean algebra** in a digital circuit. The components of a digital circuit are called **logic gates**.

Truth tables provide the rules for the Boolean operators.

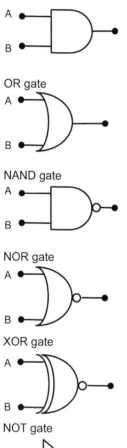

AND gate

A	B	A AND B	A•B
0	0	0	Output true
0	1	0	only if both A and B
1	0	0	are true
1	1	1	**AND**

OR gate

A	B	A OR B	A+B
0	0	0	Output true
0	1	1	if either A or B are
1	0	1	true
1	1	1	**OR**

NAND gate

A	B	A NAND B	
0	0	1	Output false if
0	1	1	both A and B are
1	0	1	true
1	1	0	**NAND**

NOR gate

A	B	A NOR B	
0	0	1	Output false if either
0	1	0	A or B is true.
1	0	0	
1	1	0	**NOR**

XOR gate

A	B	A XOR B	
0	0	0	True if either A or B
0	1	1	is true but not both
1	0	1	together.
1	1	0	**XOR**

NOT gate

A	NOT A	
0	1	True if A is false
1	0	False if A is true **NOT**

11.2 Logic gate characteristics

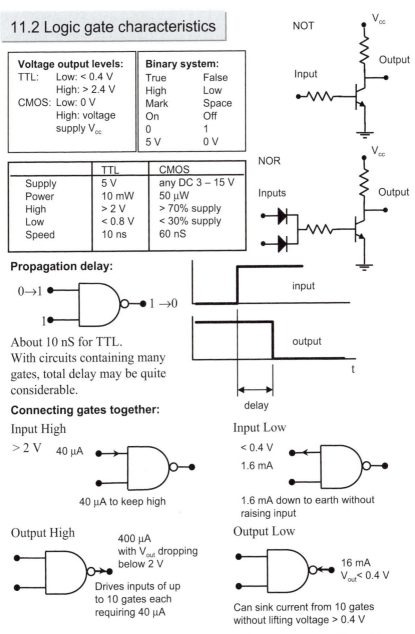

	TTL	CMOS
Supply	5 V	any DC 3 – 15 V
Power	10 mW	50 μW
High	> 2 V	> 70% supply
Low	< 0.8 V	< 30% supply
Speed	10 ns	60 nS

Voltage output levels:
TTL: Low: < 0.4 V
       High: > 2.4 V
CMOS: Low: 0 V
       High: voltage
       supply V_{cc}

Binary system:

True	False
High	Low
Mark	Space
On	Off
0	1
5 V	0 V

NOT

NOR

Input

Inputs

Output

Output

V_{cc}

V_{cc}

Propagation delay:

$0 \rightarrow 1$

1

$1 \rightarrow 0$

input

output

t

delay

About 10 nS for TTL.
With circuits containing many
gates, total delay may be quite
considerable.

Connecting gates together:

Input High

> 2 V   40 μA

40 μA to keep high

Input Low

< 0.4 V
1.6 mA

1.6 mA down to earth without
raising input

Output High

400 μA
with V_{out} dropping
below 2 V

Drives inputs of up
to 10 gates each
requiring 40 μA

Output Low

16 mA
V_{out}< 0.4 V

Can sink current from 10 gates
without lifting voltage > 0.4 V

Fan-out - number of inputs that can be driven by one output. TTL: = 10
CMOS: = 50

11.5 Digital logic circuits

Boolean algebra can be implemented using digital electronic circuits using combinations of **logic gates**.

e.g. a combination of NAND gates gives a logical XOR function.

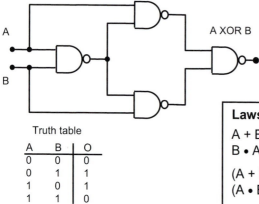

Truth table

A	B	O
0	0	0
0	1	1
1	0	1
1	1	0

In the circuit below, the XOR function is used to add binary digits A and B. The AND gate indicates whether or not there is a **carry** bit. This circuit is a **half adder**.

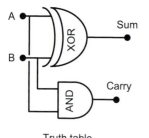

Truth table

A	B	S	C
0	0	0	0
0	1	1	0
1	0	1	0
1	1	0	1

Laws of Boolean algebra

$A + B = B + A$
$B \cdot A = A \cdot B$

$(A + B) + C = A + (B + C)$
$(A \cdot B) \cdot C = A \cdot (B \cdot C)$

$A + AB = A \cdot (1 + B) = A$
$A \cdot (A + B) = A$

$A \cdot (B + C) = A \cdot B + A \cdot C$
$A + (B \cdot C) = (A + B) \cdot (A + C)$

$A + A = A$
$A \cdot A = A$

$A \cdot \overline{A} = 0$
$A + \overline{A} = 1$
$\overline{\overline{A}} = A$
$0 + A = A$
$1 \cdot A = A$
$1 + A = 1$
$0 \cdot A = 0$

$A + \overline{A} \cdot B = A + B$
$A \cdot (\overline{A} + B) = A \cdot B$

De Morgan's theorem

$\overline{(A + B)} = \overline{A} \cdot \overline{B}$
$\overline{(A \cdot B)} = \overline{A} + \overline{B}$

11.6 Boolean logic examples

1. Associative law   (A + B) + C = A + (B + C)

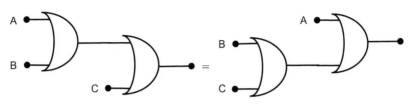

2. **De Morgan's theorem**

$$\overline{(A + B)} = \overline{A} \cdot \overline{B}$$
$$\overline{(A \cdot B)} = \overline{A} + \overline{B}$$

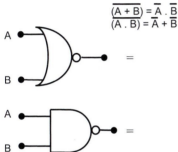
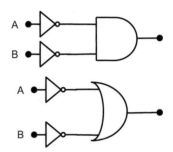

3. Example   Out = A . C + B . $\overline{C}$ + $\overline{A}$. B . C

OR └──▶ AND

1. Work on the OR's first   A . C

          B . $\overline{C}$

          $\overline{A}$. B . C

2. Add on AND's

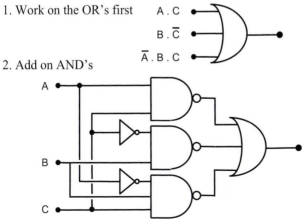

11.7 Logic circuit analysis

Consider now a more systematic approach to formulating logic circuits.

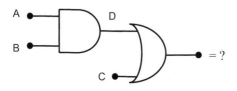

1. Draw up truth table

A	B	C	D = AB	Out = D+C	Min terms
0	0	0	0	0	$\overline{A} \cdot \overline{B} \cdot \overline{C}$
0	0	1	0	1	$\overline{A} \cdot \overline{B} \cdot C$
0	1	0	0	0	$\overline{A} \cdot B \cdot \overline{C}$
0	1	1	0	1	$\overline{A} \cdot B \cdot C$
1	0	0	0	0	$A \cdot \overline{B} \cdot \overline{C}$
1	0	1	0	1	$A \cdot \overline{B} \cdot C$
1	1	0	1	1	$A \cdot B \cdot \overline{C}$
1	1	1	1	1	$A \cdot B \cdot C$

2. AND inputs together with 0's negated - min terms

3. Circle min terms showing a 1

4. OR circled min terms to form Boolean expression

$$Out = \overline{A} \cdot \overline{B} \cdot C + \overline{A} \cdot B \cdot C + A \cdot \overline{B} \cdot C + A \cdot B \cdot \overline{C} + A \cdot B \cdot C$$
$$= \overline{A} \cdot C \cdot \left(\overline{B} + B\right) + A \cdot B \cdot \overline{C} + A \cdot C \cdot \left(\overline{B} + B\right)$$

but $\overline{A} \cdot C \cdot \left(\overline{B} + B\right) = \overline{A} \cdot C$ and $A \cdot C \cdot \left(\overline{B} + B\right) = A \cdot C$

thus, $\overline{A} \cdot C + A \cdot C = C \cdot \left(\overline{A} + A\right)$
$$= C$$

5. Then simply using Boolean algebra

therefore $Out = C + A \cdot B \cdot \overline{C}$

but $X + \overline{X} \cdot Y = X + Y$

thus : $\boxed{Out = A \cdot B + C}$ = Simplified expression

11.8 Karnaugh map

The Karnaugh map is a graphical method of analysing logic circuits.

1. Draw up truth table and circle min terms (using previous example)

A	B	C	Out = D+C	Min terms
0	0	0	0	$\overline{A} \cdot \overline{B} \cdot \overline{C}$
0	0	1	1	$\overline{A} \cdot \overline{B} \cdot C$
0	1	0	0	$\overline{A} \cdot B \cdot \overline{C}$
0	1	1	1	$\overline{A} \cdot B \cdot C$
1	0	0	0	$A \cdot \overline{B} \cdot \overline{C}$
1	0	1	1	$A \cdot \overline{B} \cdot C$
1	1	0	1	$A \cdot B \cdot \overline{C}$
1	1	1	1	$A \cdot B \cdot C$

This example is a 3 input system

2. Construct map as follows:
   (a) arrange rows and columns with every combination of input, changing only one variable at a time;

	C	$\overline{C}$
AB	1	1
$\overline{A}B$	1	
$\overline{A}\overline{B}$	1	
A$\overline{B}$	1	

Right	Wrong
AB	AB
$\overline{A}B$	$\overline{AB}$
$\overline{AB}$	$A\overline{B}$
$A\overline{B}$	$\overline{A}B$

   (b) put 1's in boxes corresponding to circled min terms in truth table;

   (c) draw boxes around groups of 1's. Boxes can only go vertically and horizontally. Boxes can also wrap around. Can only box even groups (powers of 2). Boxes of 3, 5 and 6 etc are not permitted;

   (d) group contents of boxes by products and join boxes together with sums (**Note: these are not yet Boolean AND's and OR's, we are simply following a procedure that will lead to a Boolean expression**);

   Output = $ABC\,\overline{A}BC\,\overline{A}\overline{B}C\,A\overline{B}C + ABC\,AB\overline{C}$

   (e) cancel terms of opposite sign (i.e. remove them from the expression);

   Output = $\cancel{A}\cancel{B}C\,\cancel{A}\cancel{B}C\,\cancel{A}\cancel{B}C\,\cancel{A}\overline{B}C + AB\cancel{C}\,AB\cancel{\overline{C}}$

   $= C + AB$

   (f) now replace products with "AND" and sums with "OR" to obtain final logical expression.

   Output = $C + A \cdot B$

Consider this 4 input system:

	CD	$\overline{C}D$	$\overline{C}\overline{D}$	$C\overline{D}$
AB		1	1	
$\overline{A}B$		1	1	
$\overline{A}\overline{B}$				
$A\overline{B}$				

Output $= AB\overline{C}D \cdot AB\overline{C}\overline{D} \cdot \overline{A}B\overline{C}D \cdot \overline{A}B\overline{C}\overline{D}$

$= B \cdot \overline{C}$

Groups of 4 eliminate 2 variables.
Groups of 8 eliminate 3 variables.

	CD	$\overline{C}D$	$\overline{C}\overline{D}$	$C\overline{D}$
AB	1	1		
$\overline{A}B$				
$\overline{A}\overline{B}$				
$A\overline{B}$	1	1		

Boxes can wrap around

Output $= ABCD \cdot AB\overline{C}D \cdot A\overline{B}CD \cdot A\overline{B}\overline{C}D$

$= A \cdot D$

Remember: you can only group 2's, 4's, 8's, etc. Groups can be made from wraps around from top and bottom and sides. You can also group the four corners into one group of four. A "1" on its own cannot be grouped and the associated min term must be + with the other grouped expressions.

Don't care states

Sometimes, it doesn't matter if there is a zero or a 1 in a logic circuit. On a Karnaugh map, an "X" is inserted in the position to indicate "don't care". X's may be grouped if desired if this leads to minimisation.

11.9 Flip-flops

Flip-flops can be used to represent binary numbers. An RS flip-flop is a digital circuit which is stable in one of two states – **set** or **reset**. Such a circuit can be made using NAND gates. A truth or **action table** summarises the action of flip-flop. The voltage of one of the outputs can be used to represent or store a binary digit since it can be either voltage high (logic 1) or low (logic 0) and will remain at that setting until signals on the input, which only last for a short time, set or reset the outputs.

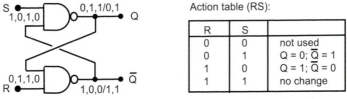

Action table (RS):

R	S	
0	0	not used
0	1	$Q = 0; \overline{Q} = 1$
1	0	$Q = 1; \overline{Q} = 0$
1	1	no change

S and R are normally held at 1 and the outputs remain constant in any one of two states. Either $Q = 0$; $\overline{Q} = 1$, or $Q = 1$; $\overline{Q} = 0$. An input sequence of 101 at S (with R = 1) ensures that $Q = 1$; $\overline{Q} = 0$. An input sequence of 101 at R (with S = 1) ensures that $Q = 0$; $\overline{Q} = 1$. In normal circuit design, the condition $S = R = 0$ should not be allowed since $Q = \overline{Q} = 1$ is not very useful. For a flip-flop using NOR gates, the inputs $S = R = 1$ result in both outputs being at logic 0, again, not a useful condition for a circuit whose main feature is to have the outputs opposite to each other.

Clocked flip-flops are used in computers and are set on receipt of a timing or clock pulse.

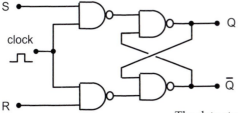

Data at terminal S gets transferred to Q on the clock pulse and remains at Q even if the signal at S disappears and the clock goes low.

Action table (clocked RS):

R	S	
0	0	no change
0	1	$Q = 1; \overline{Q} = 0$
1	0	$Q = 0; \overline{Q} = 1$
1	1	not used

The data stays at Q because when the clock pulse goes low, the flip-flop circuits *within the chip* are at $S = R = 1$ (due to the NAND gates on the clock stage). Only when the clock goes high do the flip-flops react to the logic signals at D on the latch.

Note: this action table is different to the ordinary NAND flip-flop. Here, $R = S = 1$ is the "not used" state.

11.10 D-latch

A **latch** is a device which holds the data that appears on its input terminals. A memory cell in the microcomputer system is a latch. Typically, signals destined for storage in memory cells appear on the data bus momentarily and then disappear. The timing of the signals is regulated by the internal **clock** which runs at speeds typically in the MHz range. The decoding circuitry determines which buffer is to be activated. The activated buffer in turn connects the latch input terminals to the data bus. The signals on the data bus are transferred through buffers to the latch circuit which stores the signals on its output terminals.

A latch circuit can be implemented using a series of RS **flip-flops**. In this figure, the 4 bit data at D_3 to D_0 is transferred to Q on the clock pulse. When a bit D is logic 1, $S = 1$ and $R = 0$ and the output Q becomes 1. When D is logic 0, $S = 0$ and $R = 1$ and the output $D = 0$.

An **octal latch** has 8 inputs and 8 outputs. The data latch enable (DLE) pin, when set high, copies the voltage levels on the input pins to the corresponding output. The latch circuitry retains the signals on the output pins even if the input signals disappear and DLE goes low. It is important that DLE is set when data appears on the input. DLE is typically timed to go high when data appears on the data bus. The clock signals are used to synchronise this timing.

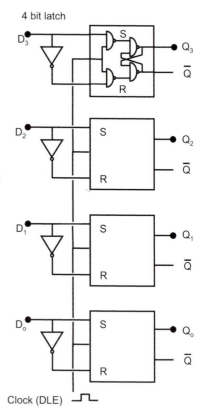

4 bit latch

Clock (DLE)

11.11 J-K master-slave flip-flop

Two clocked RS flip-flops in tandem constitute what is called a J-K or "master-slave" flip-flop. The two inputs to the device as a whole are labelled J and K for convenience and to avoid confusion with the terms R and S. The advantage of this arrangement is that the output responds to the state of the input only on the falling of the clock pulse and is not affected by jitter on the input when the clock is high.

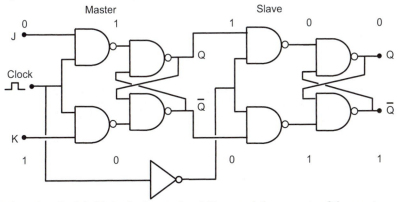

When the clock is high, the inputs J and K control the outputs of the master according to the action table for a regular RS flip-flop. The output of the slave is not affected since its clock is low (via the inverter). When the clock pulse goes low, the outputs on the master are isolated from any changes in the input, and the slave flip-flop is now activated and the outputs respond to the output levels of the master.

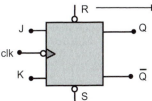

The terminals "S" and "R" are "set" and "clear" (active low) which are *not* shown on the above circuit diagram.

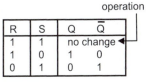

Normal operation

R	S	Q	Q̄
1	1	no change	
1	0	1	0
0	1	0	1

R & S override J,K. R = S = 0 should not be used.

The net effect is for the inputs J and K to be read on the high part of the clock cycle and the output only responds to this input when the clock falls - hence the inversion bubble on clk.

Action table:

J	K	Clock pulse 1 to 0
0	0	no change, $Q_{n+1}=Q_n$
0	1	$Q_{n+1} = 0$ (RESET)
1	0	$Q_{n+1} = 1$ (SET)
1	1	$Q_{n+1}= \bar{Q}_n$ toggle

11.12 J-K flip-flop examples

1. If the inputs of a JK flip-flop are held at 1, then the flip-flop is placed in **toggle mode**.

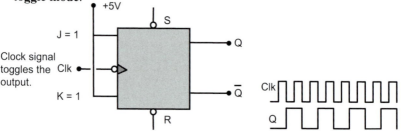

For a square wave input at the clock, the output Q is a square wave but half the frequency. The toggle mode connection is often called a **divide by two** circuit for this reason.

2. JK flip-flop as a D-type latch: clock signal transfers data from D to Q.

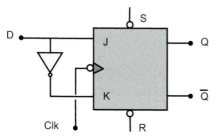

The **asynchronous inputs** R and S are normally held at 1 (depending on the IC being used). Setting S or R to 0 either sets (Q = 1) or resets (Q = 0) the flip-flop and overrides signals on J and K (with S and R active low).

3. A chain of J-K flip-flops in toggle mode

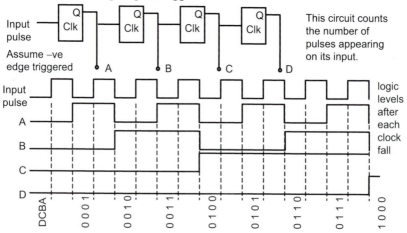

This circuit counts the number of pulses appearing on its input.

11.13 Monostable multivibrator

Monostable **multivibrators** are used as "wave shaping" circuits. For example, it may be necessary to apply an output pulse which remains high for a pre-determined time on receipt of an input trigger pulse.

Consider the following circuit:

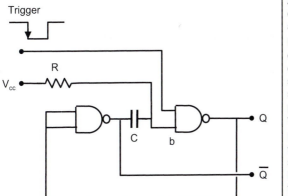

Astable: the outputs oscillate between 0 and 1 continuously.

Monostable: the output terminals have one stable state and switch to the opposite state for a short time period before returning to the stable state.

Bistable: the output terminals have two stable states, and can be triggered from one to the other.

In this circuit, the trigger level is normally high. Thus, the output Q is low since this level is NANDed with V_{cc} through R_c. With Q low, the output $\overline{Q}$ is high due to the inverting action of the first NAND gate. Since both sides of the capacitor are high, then there is no charge build-up on the plates and the circuit is stable in this condition.

When the trigger level falls momentarily, Q goes high which sends $\overline{Q}$ low. The voltage level at b drops to zero and rises back to V_{cc} as the capacitor charges according to a time constant which depends on the value of R and C.

When the capacitor charges up fully voltage at b reaches V_{cc} and, since the trigger has already returned high, the output Q goes high and the circuit resumes its stable condition.

This is a single-shot **monostable multivibrator**. Variations on this circuitry allows the circuit to ignore any further triggers until the output goes back high (non-retriggerable).

11.14 555 Timer

A popular choice for timing applications is the **555 timer** IC.

Monostable operation

The 555 contains two comparators whose reference voltages are set at 2/3 and 1/3 V_{cc} by internal resistors. The trigger terminal is usually kept high by an external circuit. When the trigger falls to 1/3 V_{cc}, Comparator #1 sets the flip-flop and the output $\overline{Q}$ goes low. The output of the IC goes high via the inverter. When $\overline{Q}$ goes low, the transistor is turned off and the capacitor begins to charge. When the voltage at the capacitor reaches 2/3 V_{cc}, Comparator #2 toggles the flip-flop and the transistor is turned on by $\overline{Q}$, thus discharging the capacitor to earth. Thus, a low-going pulse on the trigger causes a high pulse on the output which stays high according to the values of R and C in the external circuit.

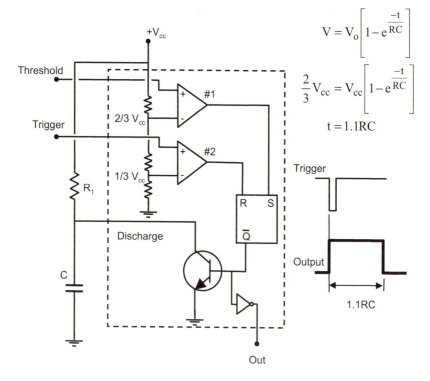

$$V = V_0 \left[1 - e^{\frac{-t}{RC}} \right]$$

$$\frac{2}{3} V_{cc} = V_{cc} \left[1 - e^{\frac{-t}{RC}} \right]$$

$$t = 1.1RC$$

Astable operation

The 555 timer IC can be wired for astable operation. It is most often used as an astable timer.

> An astable device has no stable state. Its outputs oscillate between set and reset condition at a fixed frequency. External components (e.g. resistor and capacitor) are usually used to set the frequency of oscillation.

The 555 contains two comparators whose reference voltages are set at 2/3 and 1/3 V_{cc} by internal resistors. The capacitor C charges through R_1 and R_2. When the voltage at threshold reaches 2/3 V_{cc} the flip-flop is set by comparator #1. In this flip-flop, $\overline{Q}$ goes high which turns on the transistor and discharges the capacitor through R_2. When the voltage at the capacitor falls to 1/3 V_{cc}, comparator #2 resets the flip-flop and the transistor is turned off and the capacitor begins to charge again.

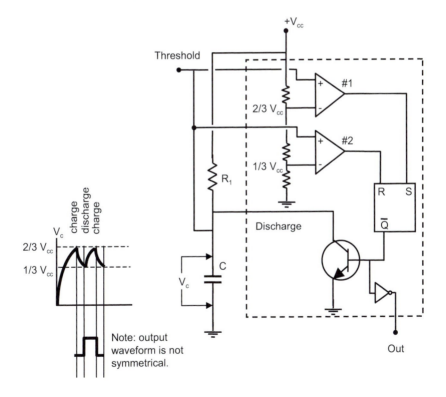

Review questions

1. Show how an inverter can be made using a (a) NOR gate, (b) NAND gate.

2. Derive an expression for the XOR function using AND and OR expressions.

3. Design a logic circuit which implements the AND function but using NOR gates only.

4. Design a logic circuit which implements the XOR function but using OR and NOR gates only.

5. Design a logic circuit which implements the XOR function but using AND and NAND gates only.

6. Express the logic operations of the circuit below in algebraic form.

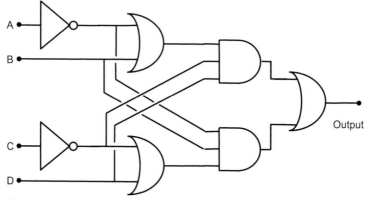

7. Given two TTL packages, one a hex inverter (containing 6 independent NOT gates) and the other a quad 2-input NAND (containing 4 independent, 2 input NAND gates), design a single 4 input gate system to give: $\overline{A + B + C + D}$

8. For a three input signal A,B, and C, the exclusive OR operation can be written:

$$\left(A + B + C\right)\left(\overline{A.B.C}\right)$$

Design a logic circuit to perform this operation using two and three-input NAND gates only. No more than 7 gates should be used.

9. Draw Karnaugh maps for the following logic operations and indicate the simplest form of the circuit using NAND gates.

(a) $ABC\overline{D} + ABCD + A\overline{B}C\overline{D} +$

$\overline{AB}CD + \overline{A}BCD + A\overline{BCD}$

(b) $\overline{ABCD} + \overline{ABCD} + \overline{ABCD} + \overline{ABCD} +$

$\overline{ABCD} + A\overline{BCD} + AB\overline{CD} + \overline{ABCD}$

10. Design a logic circuit which gives an output of 1 whenever a 4 bit binary input is an even number (including zero).

11. Construct a Karnaugh map and derive a Boolean expression to implement the following truth table.

A	B	C	O
0	0	0	0
0	0	1	0
0	1	0	0
0	1	1	1
1	0	0	0
1	0	1	1
1	1	0	1
1	1	1	1

12. Two 2 bit binary numbers A and B are represented by the bits A0, A1 and B0, B1. Design a simple logic circuit with 3 outputs X,Y and Z, such that X = 1 when A > B, Y = 1 when A = B and Z = 1 when A < B. Use NAND and NOR gates only.

13. Sketch the output Q waveform resulting from the inputs to the RS
flip-flop which operates as a NAND gate (assuming Q starts low):

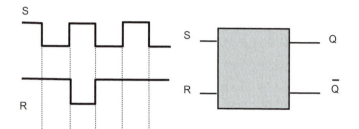

14. A +ve edge triggered clocked RS flip-flop receives the following
signals. Sketch the output Q.

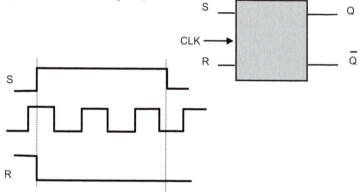

15. Sketch the Q output for the input as shown below to a D latch.

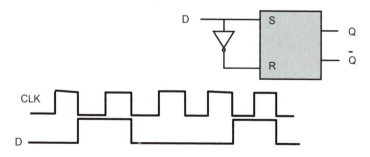

12. Operational amplifiers

Summary

$$A_o = \frac{V_{out}}{V_d}$$ Open loop gain

$$A_o = \frac{V_{out}}{V_d}$$ Differential gain

$$A_C = \frac{A_o}{1 + \beta A_o}$$ Negative feedback

$$A_C = -\frac{R_2}{R_1} = \frac{1}{\beta}$$ Inverting amplifier

$$A_c = \frac{R_2 + R_1}{R_1}$$ Non-inverting amplifier

$$A_{cm} = \frac{R_e}{2R_c}$$ Common mode gain

$$A_v = -\left(\frac{2R_a}{R} + 1\right)\frac{R_2}{R_1}\left(V_1 - V_2\right)$$ Instrumentation amplifier

12.1 Differential amplifier

A **differential amplifier** has two inputs. The output voltage is measured with respect to earth. The amplifier can be used to amplify both DC and AC. The circuit amplifies the difference between the voltages on the two inputs.

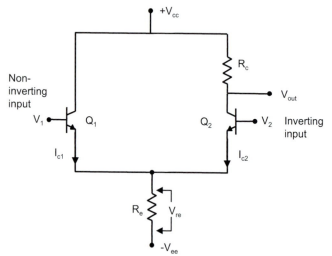

1. When V_1 increases (by just a little), I_{c1} increases and therefore more current flows through R_e thus causing V_{re} to increase.

2. Increasing V_{re} means a decrease in V_{be} of Q_2 and this leads to a decrease in I_{b2} and hence a decrease in I_{c2}.

3. A decrease in I_{c2} means an increase in V_{out} (w.r.t. earth) since there is less voltage drop across R_c.

> V_1 is a **non-inverting input**. An increase in V_1 results in an increase in V_{out}.

4. When V_2 increases, I_{c2} increases and hence there is a decrease in V_{ce} and a larger voltage drop across R_c. Thus, there is a decrease in V_{out} (w.r.t. earth).

> V_2 is an **inverting input**. An increase in V_2 results in a decrease in V_{out}.

A useful connection is to tie the base of both transistors to ground through resistors R_b.

The product $I_b R_b$ is small because I_b is small. Therefore, the voltage at the base is approximately 0 V. Thus, the potential at "e" = -0.7 V and thus:

$$V_{re} = 15 - 0.7$$

$$\approx -V_{ee}$$

The **tail current** I_T is found from:

$$I_T = \frac{V_{ee}}{R_e}$$

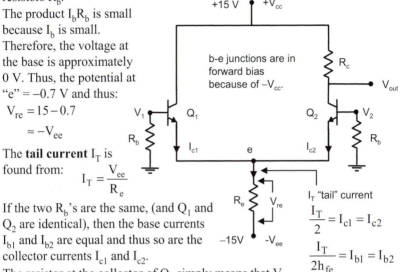

If the two R_b's are the same, (and Q_1 and Q_2 are identical), then the base currents I_{b1} and I_{b2} are equal and thus so are the collector currents I_{c1} and I_{c2}.

$$\frac{I_T}{2} = I_{c1} = I_{c2}$$

$$\frac{I_T}{2h_{fe}} = I_{b1} = I_{b2}$$

The resistor at the collector of Q_2 simply means that V_{ce} for Q_2 does not equal V_{ce} for Q_1.

The **input offset current** is the difference $I_{off} = I_{b1} - I_{b2} \approx 10\text{ nA}$ and arises due to differences in transistors (i.e. differences in IV curves).

The **input bias current** is, by definition, the average of the two input bias currents.

$$I_{bias} = \frac{I_{b1} + I_{b2}}{2}$$

The **output offset voltage** arises due to differences in transistors. The ideal output voltage is:

$$V_{out} = V_{cc} - \frac{I_T}{2} R_c$$

However, if the transistors are not identical, then V_{be1} does not equal V_{be2} and thus there is a change in the output given by the difference in gain A_d:

$$= \frac{V_{cc}}{2}$$

Assumes:
$$I_T = I_{c1} + I_{c2}$$
$$I_{c1} = I_{c2}$$

$$\Delta V_{out} = A_d (\Delta V_{be})$$

Because I_b changes, so does I_c and hence $I_T \neq \frac{I_{c1}}{2} + \frac{I_{c2}}{2}$

$$V_{b1} = I_{b1} R_b$$
$$V_{b2} = I_{b2} R_b$$

$$\longrightarrow \Delta V_{be} = I_{b1} R_b - I_{b2} R_b$$

The **offset null** is a small DC signal applied to one transistor input to eliminate output offset voltage.

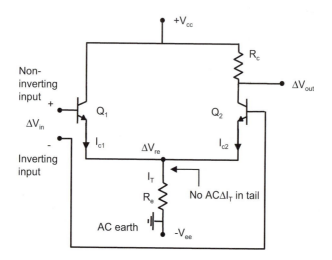

An increase at +ve input gives a decrease I_{c2} and an increase in V_{out}. A decrease in the ⁻ve input gives a decrease in I_{c2} and increase in V_{out}. The output thus depends on the magnitude of the difference in voltage on the inputs.

I_T is the total of collector currents: $I_T = I_{c1} + I_{c2}$

If I_T is more or less constant, then $\Delta I_T = 0$ under ideal conditions, thus:

$$\Delta I_{c1} + \Delta I_{c2} = 0$$

The **voltage gain** is found from: $\Delta V_{in}^+ = \Delta I_{b1} h_{ie1} + \Delta V_{re}$ Floating level, R_e acts like an

$$\Delta V_{in}^- = \Delta I_{b2} h_{ie2} + \Delta V_{re}$$ open circuit.

Let $h_{ie1} = h_{ie2}$, then: $\Delta V_{in} = \left(\Delta V_{in}^+ - \Delta V_{in}^- \right)$

$$= h_{ie} \left(\Delta I_{b1} - \Delta I_{b2} \right)$$

$$= 2 h_{ie} \Delta I_b \qquad \text{letting } \Delta I_{b1} = -\Delta I_{b2}$$

$$= 2 h_{ie} \frac{\Delta I_c}{h_{fe}}$$

$$\Delta V_{out} = -\Delta I_{c2} R_c$$

$$A_d = \frac{\Delta V_{out}}{\Delta V_{in}} = -\frac{h_{fe} R_c}{2 h_{ie}}$$

Input resistance

12.2 Operational amplifier

An **operational amplifier** is a differential amplifier with a high AC and DC gain.

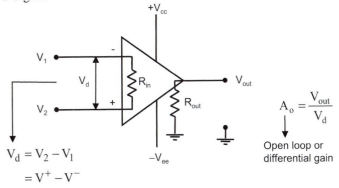

$$V_d = V_2 - V_1$$
$$= V^+ - V^-$$

$$A_o = \frac{V_{out}}{V_d}$$

Open loop or differential gain

Ideal characteristics:

1. A_o is very high ($\approx 10^6$) which gives stability with feedback components
2. infinite bandwidth - A_o is constant over a wide range of frequencies
3. R_{in} is very high
4. R_{out} is close to zero
5. minimal drift and low noise
6. maintains a constant phase relationship between input and output.

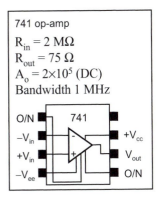

741 op-amp

$R_{in} = 2\ M\Omega$
$R_{out} = 75\ \Omega$
$A_o = 2 \times 10^5$ (DC)
Bandwidth 1 MHz

12.3 Feedback

Consider this block diagram of a difference amplifier:

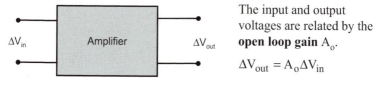

The input and output voltages are related by the **open loop gain** A_o.

$$\Delta V_{out} = A_o \Delta V_{in}$$

Now consider the same amplifier with a factor β **feedback** which is subtracted from the input signal:

$\Delta V_{in} = \Delta V_s - \beta \Delta V_{out}$

ΔV_s is signal to be amplified
ΔV_{in} is signal to amplifier

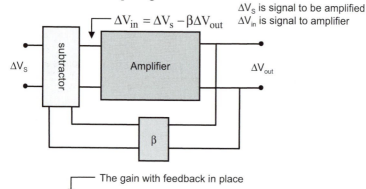

The gain with feedback in place

The **closed loop gain** is given by:

$$A_C = \frac{\Delta V_{out}}{\Delta V_S}$$

thus

$$\frac{\Delta V_{in}}{\Delta V_{out}} = \frac{\Delta V_S - \beta \Delta V_{out}}{\Delta V_{out}}$$

$$= \frac{\Delta V_S}{\Delta V_{out}} - \beta$$

$$\frac{1}{A_o} = \frac{1}{A_C} - \beta$$

$$\boxed{A_C = \frac{A_o}{1 + \beta A_o}}$$

Negative feedback reduces the voltage gain but as we shall see, offers other benefits.

Note: when A_o is very high, then the closed loop gain A_C approaches:

$$\boxed{A_C = \frac{1}{\beta}}$$

12.4 Inverting amplifier

When resistors are connected to allow feedback, the gain of the circuit is reduced and is termed the **closed loop gain** A_c. Shown below is an op-amp with one terminal grounded and an input signal (AC or DC) applied (via R_1) to the inverting input. This is a **single-ended** mode of operation.

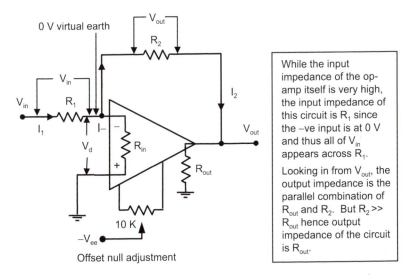

While the input impedance of the op-amp itself is very high, the input impedance of this circuit is R_1 since the –ve input is at 0 V and thus all of V_{in} appears across R_1.

Looking in from V_{out}, the output impedance is the parallel combination of R_{out} and R_2. But $R_2 \gg R_{out}$ hence output impedance of the circuit is R_{out}.

The input resistance R_{in} within the op-amp is typically very high so I^- is very small. Thus, $I_1 = I_2$ and there is negligible voltage drop across R_{in}. Hence, the voltage at the inverting input = 0 V but actually isolated from earth. It is thus called a **virtual earth**.

The current I_1 flows through R_1 and, since I^- is negligible, I_1 must then go through R_2 and to earth through the low impedance R_{out}. Since the ⁻ve input is at 0 V (virtual earth) it follows that the potential at V_{out} must be < 0, hence the term: **inverting amplifier**.

$$V_{out} = -I_2 R_2$$
$$V_{in} = I_1 R_1$$
$$\frac{V_{out}}{V_{in}} = -\frac{R_2}{R_1} \qquad I_1 = I_2$$

$$\boxed{A_C = -\frac{R_2}{R_1} = \frac{1}{\beta}}$$ A_c is the closed loop gain of the inverting amplifier and depends only on the value of external resistors.

12.5 Non-inverting amplifier

With a little rearrangement, the inverting amplifier can be connected to form a non-inverting amplifier.

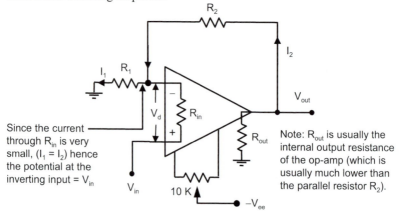

Since the current through R_{in} is very small, ($I_1 = I_2$) hence the potential at the inverting input = V_{in}

Note: R_{out} is usually the internal output resistance of the op-amp (which is usually much lower than the parallel resistor R_2).

Since the potential at -ve input is at V_{in}, then current I flows from −ve to ground through R_1. Also, since this is a non-inverting amplifier, $V_{out} > V_{in}$ so current I_2 flows from V_{out} towards −ve input through R_2.

$$V_{in} = IR_1$$
$$V_{out} = IR_2 + V_{in}$$
$$\frac{V_{out}}{V_{in}} = \frac{IR_2 + IR_1}{IR_1}$$

$$\boxed{A_c = \frac{R_2 + R_1}{R_1}}$$

Note: the negative feedback ratio is:

$$\beta = \frac{R_2 + R_1}{R_1}$$

A_c is the closed loop gain non-inverting amplifier and depends only on external resistors.

In this case, the input signal is applied directly to the +ve input on the op-amp and hence sees the high R_{in} associated with the circuitry of the op-amp itself. The output impedance is dominated by the relatively low output resistance of the op-amp but can be reduced by the presence of R_2 in the circuit.

Note: if the currents I^+ and I^- are "negligible", and that the voltage at both inputs is the same (0 V for inverting, V_{in} for non-inverting) then you might wonder how does the whole thing work? The amplifier works because there is (i) an infinite open loop gain, and (ii) feedback of the output back to the input. The combination of these enables the closed loop gain to be dependent on the external resistors only.

12.6 Offset

With both inputs at zero, V_{out} does not usually equal zero due to V_{be} of
each input transistor usually being different.

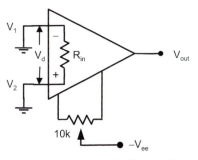

Offset null allows adjustment to V_d to
make $V_{out} = 0$ when $V_1 = V_2$. The
amount by which V_d is changed is called
the **input offset voltage** V_{os}

The first stage of an op-amp is a differential amplifier. The input transistors
require input bias currents I^- and I^+. Thus a path to ground is required (cannot
have any of the inputs "open" or floating).

This circuit would
not work because
there is no path
to ground for I^+.

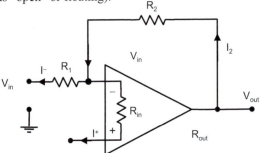

The **input bias current** I_{bias} is the average of I^+ and I^-.

Due to differences in transistors, I^- and I^+ are usually not the same. The
difference is called the **input offset current** and is usually an order of
magnitude less than the average input bias current.

Typical values: I_{bias} = 80 nA, I_{off} = 20nA.

The input offset voltage, input bias current and input offset current all
contribute to an undesirable DC voltage at the output terminal.

12.7 Op-amp bias

Consider an inverting amplifier. If both inputs are grounded, then let us assume that any output V_{out} arises solely from I_{off} (so that we can see what effect I_{off} has).

Note: in these types of circuits, current into the Op-Amp itself (I^+ and I^-) flows from earth into the +ve and −ve inputs. How can this be? Because of the presence of $-V_{cc}$ which is below earth potential. Thus, the transistors within the op-amp are in forward bias (assuming npn).

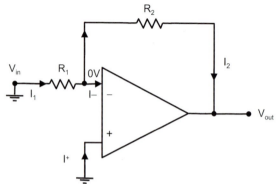

Consider the resistance to earth as seen by both inputs. For the −ve input, it is evident that R_1 and R_2, under these circumstances, are effectively in parallel but there is no corresponding resistance to earth for the +ve input. This would lead to an input voltage differential as the input bias currents would be drawn through different resistances. To avoid this, an additional balancing resistor R_x has to be inserted at the +ve input:

$$R_x = \frac{R_1 R_2}{R_1 + R_2}$$

R_x has no effect on the closed loop gain.

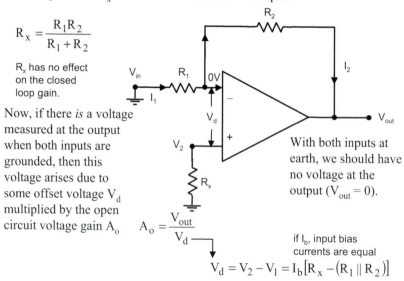

Now, if there *is* a voltage measured at the output when both inputs are grounded, then this voltage arises due to some offset voltage V_d multiplied by the open circuit voltage gain A_o

$$A_o = \frac{V_{out}}{V_d}$$

With both inputs at earth, we should have no voltage at the output ($V_{out} = 0$).

if I_b, input bias currents are equal

$$V_d = V_2 - V_1 = I_b \left[R_x - \left(R_1 \parallel R_2 \right) \right]$$

12.8 Common mode gain

One advantage of a differential amplifier is that the output ideally equals zero when the same signal appears on the inputs. This is useful for amplifying a signal which contains "noise". If the signal + noise is applied to one input, and the noise alone is applied to the other input, then the differential amplifier will provide an amplified signal and the noise will be rejected (since it is common to both inputs).

$$\Delta V_{in1} = \Delta I_b h_{ie} + \Delta V_{re}$$

$$\Delta V_{in2} = \Delta V_{in1}$$

$$= \Delta I_{b2} h_{ie} + \Delta V_{re}$$

$$\Delta V_{in1} + \Delta V_{in2} = h_{ie}\left(\Delta I_{b1} + \Delta I_{b2}\right) + \Delta V_{re}$$

$$= h_{ie}\frac{\Delta I_T}{h_{fe}} + 2\Delta V_{re} \quad\text{small}$$

$$= \Delta I_T\left(\frac{h_{ie}}{h_{fe1}} + 2R_e\right)$$

$$\approx \Delta I_T 2R_e$$

$$2\Delta V_{in} = 2\Delta V_{re} \quad \text{since } \Delta I_{b1} = -\Delta I_{b2}$$

$$= 2\Delta I_T R_e \frac{R_c}{R_c}$$

$$\Delta V_{in} = \frac{\Delta I_c}{2} R_c \frac{R_e}{R_c}$$

$$= \Delta V_{out}\frac{R_e}{2R_c}$$

$$A_{cm} = \frac{\Delta V_{out}}{\Delta V_{in}}$$

$$\boxed{A_{cm} = \frac{2R_c}{R_e}}$$

Common mode rejection ratio
This is a measure of quality for the response of an amplifier to common signals on the inputs. A good amplifier has a small common mode gain A_{cm}.

$$\text{CMRR} = \frac{A_o}{A_{cm}} = 20\log_{10}\frac{A_o}{A_{cm}}$$

Eg. CMRR = 100db, $A_o = 10^5$

$$100 = 20\log_{10}\frac{10^5}{A_{cm}}$$

$$A_{cm} = 10^{-3}$$

$$V_{out} = A_o V_{in} + A_{cm}V_{cm}$$

12.9 Op-amp applications

Voltage limiter

A **voltage follower** is a non-inverting amplifier with $A_c = 1$ and 100% feedback.

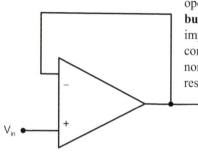

This circuit makes use of the very high open loop gain of an op-amp to make a **buffer** which has a very high input impedance. Such a buffer may be connected at the inputs to an inverting or non-inverting amplifier to present a high resistance to the signal source.

V_{out}

V_{in}

V_d normally = 0. Thus, $V_{in} = V_{out}$

Input resistance: high
Output resistance: low

$$R_{in\,new} = R_{in\,old}\left(1 + |\beta|A_o\right)$$

$$\approx R_{in\,old}\,10^5 \qquad \beta = 1$$
$$A_o = 10^5$$

Voltage limiter

It is often required for V_{out} to be between 0 V and V_{max} where $V_{max} < V_{cc}$. This can be done using a zener diode and current limiting resistor in the output. The general circuit is shown below:

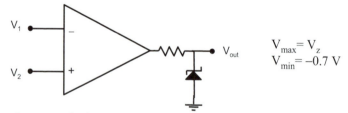

V_1

V_2

V_{out}

$$V_{max} = V_z$$
$$V_{min} = -0.7 \text{ V}$$

In another example, in the circuit below, when $V_{in} > 0$ V, V_{out} is restricted to −0.7 V. When $V_{in} < 0$, then V_{out} is restricted to a maximum of V_Z.

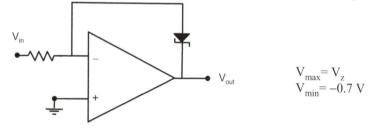

V_{in}

V_{out}

$$V_{max} = V_z$$
$$V_{min} = -0.7 \text{ V}$$

12.10 Operational adder

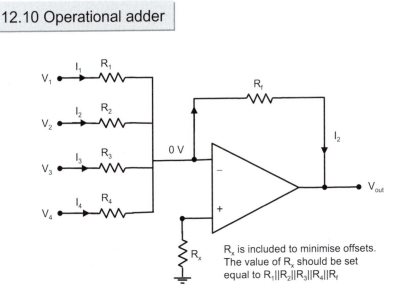

R_x is included to minimise offsets.
The value of R_x should be set
equal to $R_1||R_2||R_3||R_4||R_f$

This type of "analogue" circuit was first used to perform an adding operations hence the name **operational amplifier**. Op-amps are now used in a wide variety of other applications.

$$\frac{V_{out}}{V_{in}} = -\frac{R_f}{R_1}$$

$$V_{out} = -\left(\frac{V_{in}}{R_1}\right)R_f$$

$$V_{out} = -(i_1 + i_2 + i_3 + i_4)R_f$$

$$= -R_f\left(\frac{V_1}{R_1} + \frac{V_2}{R_2} + \frac{V_3}{R_3} + \frac{V_4}{R_4}\right)$$

Output is the weighted sum of the input voltages. If $R_1 = R_2 = R_3 = R_4 = R_f$, then the circuit simply sums the voltages on the inputs. If $R_1 = R_2 = R_3 = R_4 = R$, and $R_f = 4R$, then the circuit finds the average of the voltages on the inputs.

12.11 Comparator

A comparator circuit switches the output on a change in relative polarity on
the inputs. No feedback elements are required.

Because A_o is large, $V_o = \pm V_{cc}$ (if $V_{cc} = V_{ee}$)

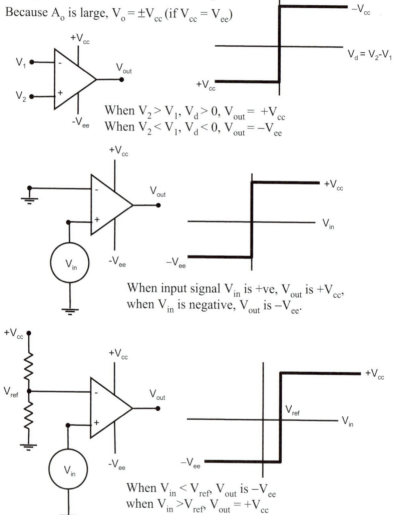

$V_d = V_2 - V_1$

When $V_2 > V_1$, $V_d > 0$, $V_{out} = +V_{cc}$
When $V_2 < V_1$, $V_d < 0$, $V_{out} = -V_{ee}$

When input signal V_{in} is +ve, V_{out} is $+V_{cc}$,
when V_{in} is negative, V_{out} is $-V_{ee}$.

When $V_{in} < V_{ref}$, V_{out} is $-V_{ee}$
when $V_{in} > V_{ref}$, $V_{out} = +V_{cc}$

A typical example is for a thermostat, where V_{in} may be the output from a
thermocouple. When the thermocouple output voltage reaches V_{ref}, V_{out} goes
positive which may operate a relay which switches off the heating element.

12.12 Schmitt trigger

Comparators do not often work as desired when the input signal contains "noise". That is, fluctuations in V_{in} may cause the output of a comparator to toggle from $+V_{cc}$ to $-V_{ee}$ as the average value of V_{in} approaches V_{ref}.

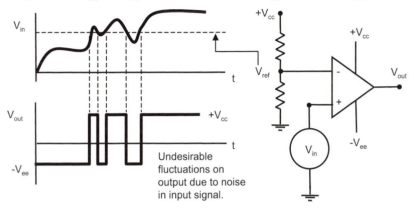

Undesirable fluctuations on output due to noise in input signal.

The **Schmitt trigger** is a variation on the basic comparator circuit designed to overcome this problem. The reference voltage is taken from a voltage divider on the output. Thus, when the output changes state, the reference voltage changes.

Negative-going Schmitt trigger: V_{in}

If V_{in} is initially less than V_{ref}, then

$$V_{out} = +V_{cc}$$

$$V_{ref} = +V_{cc} \frac{R_2}{R_1 + R_2}$$

The output will stay at $+V_{cc}$ until V_{in} goes above this value of V_{ref}. When this happens, V_{out} goes to $-V_{ee}$ and V_{ref} changes thus:

$$V_{ref} = -V_{cc} \frac{R_2}{R_1 + R_2}$$

V_{out} will remain at $-V_{ee}$ until V_{in} falls below this $-$ve value of V_{ref}. Thus, small fluctuations in V_{in} do not toggle V_{out}.

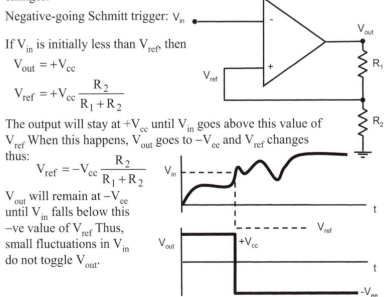

12.13 Integrator/differentiator

Integrator

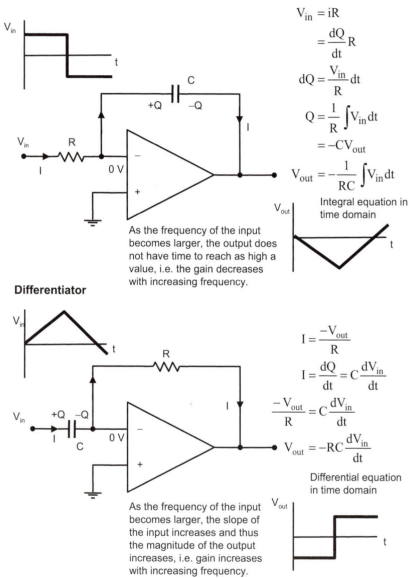

$$V_{in} = iR$$

$$= \frac{dQ}{dt} R$$

$$dQ = \frac{V_{in}}{R} dt$$

$$Q = \frac{1}{R} \int V_{in} dt$$

$$= -CV_{out}$$

$$V_{out} = -\frac{1}{RC} \int V_{in} dt$$

Integral equation in time domain

As the frequency of the input becomes larger, the output does not have time to reach as high a value, i.e. the gain decreases with increasing frequency.

Differentiator

$$I = \frac{-V_{out}}{R}$$

$$I = \frac{dQ}{dt} = C \frac{dV_{in}}{dt}$$

$$\frac{-V_{out}}{R} = C \frac{dV_{in}}{dt}$$

$$V_{out} = -RC \frac{dV_{in}}{dt}$$

Differential equation in time domain

As the frequency of the input becomes larger, the slope of the input increases and thus the magnitude of the output increases, i.e. gain increases with increasing frequency.

12.14 Instrumentation amplifier

An **instrumentation amplifier** is has a high gain and high CMRR. It is formed by using cross-coupled inputs for high CMRR and high input impedance.

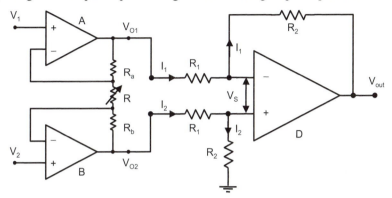

The gain of the input stage is:

$$\frac{V_{o1} - V_{o2}}{V_1 - V_2} = \frac{(R_a + R_b + R)}{R}$$

The gain of the output stage is:

$$A_d = \frac{R_2}{R_1}$$

Thus the total gain is:

$$A_v = -\frac{(R_a + R_b + R)}{R}\frac{R_2}{R_1}$$

$$= -\frac{(R_a + R_a + R)}{R}\frac{R_2}{R_1} \qquad \text{letting } R_a = R_b$$

$$A_v = -\left(\frac{2R_a}{R} + 1\right)\frac{R_2}{R_1}$$

Note that if $V_1 = V_2$, then $V_{out} = 0$ ("infinite" CMRR)

The resistors R_1 at the input to the output differential amplifier are trimmed to eliminate amplification of any common mode signal. It is usual to use the gain of the input stage to be the overall gain of the amplifier while the output stage is set to unity gain: $R_2/R_1 = 1$. The purpose of the output stage difference amplifier D is to simply reject any common mode signal.

12.15 Audio amplifier

Often in portable equipment, a dual polarity power supply is not available. However, an op-amp circuit can be configured to operate using a single +ve power supply as shown below.

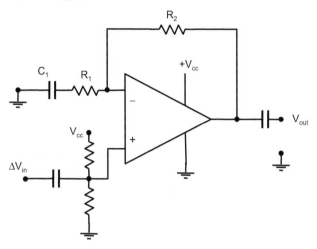

Since the $-V_{ee}$ terminal is grounded, then the output from the op-amp is always positive. Actual circuit operation however is not affected, with all inputs lifted up by $+V_{cc}/2$. The capacitor C_1 in the feedback path means that for DC, the circuit is a voltage follower with a gain of 1.

The +ve input is held at $V_{cc}/2$ by a voltage divider. Since the DC gain of the amplifier is 1, then the output also sits at $V_{cc}/2$. Coupling capacitors isolate these DC "bias" levels from the input and output devices.

Use of an op-amp as an audio amplifier allows a reasonably high gain with low distortion and good input and output impedance characteristics.

Review questions

1. A certain amplifier has an open loop gain of 250. Calculate the overall
   gain if the amplifier is constructed with 10% negative feedback.

   (Ans: 9.61)

2. If the open loop gain of an amplifier is infinite, calculate the closed-
   loop gain with 20% feedback.

   (Ans: 5)

3. A 2 mV input signal is applied to a differential amplifier on top of a
   500 mV common mode signal. If the amplifier has a differential gain of
   200, what CMRR is required for no more than 1% of the output signal
   to be contributed by the common mode input signal?

   (Ans: 86 dB)

4. Calculate the closed-loop gain and the % feedback of the circuit shown:

   (Ans: 477, 0.21%)

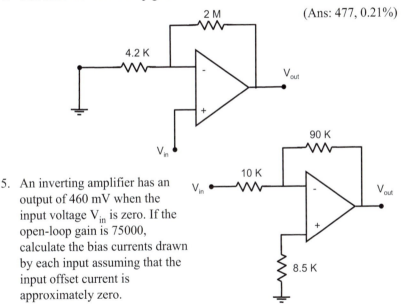

5. An inverting amplifier has an
   output of 460 mV when the
   input voltage V_{in} is zero. If the
   open-loop gain is 75000,
   calculate the bias currents drawn
   by each input assuming that the
   input offset current is
   approximately zero.

   (Ans: 12.26 nA)

6. Calculate the output voltage of the circuit
   shown and also calculate the value of R_X.

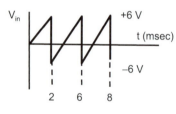

(Ans: -14.1 V, 2.7 kΩ)

7. Sketch the form of the output from the
   circuit shown below:

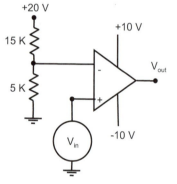

8. Determine turn on and turn off conditions for the Schmitt trigger circuit
   below (power supply is ± 10 V).

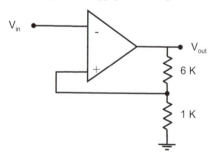

(Ans: $V_{ref} = 1.43$ V)

9. Determine which times the output goes from +10 V to −10 V in the Schmitt trigger circuit below when driven with a 1.592 kHz sine wave.

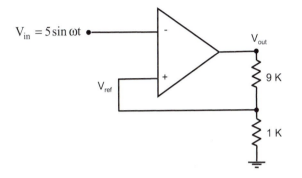

$V_{in} = 5 \sin \omega t$

V_{out}

V_{ref}

9 K

1 K

(Ans: 20.1 μs, 334 μs)

10. Sketch the output of the integrator circuit when fed with a square wave as shown:

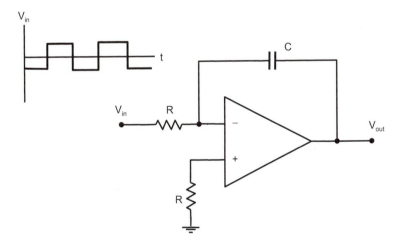

V_{in}

t

C

V_{in}

R

V_{out}

R

11. An integrator circuit has a 1 μF capacitor as the feedback element and a
10 kΩ resistor. Calculate how long it will take the output to reach the
power supply voltage −15 V if 5 mV DC is applied to the input (assume
$V_{out} = 0$ @ t = 0).

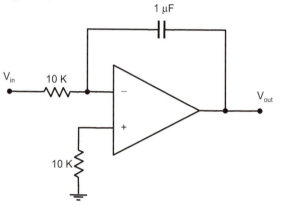

(Ans: 30 s)

12. The output of the integrator circuit shown below is 0 V at t = 0.
Calculate the output voltage at t = 2 msec, 6 msec, and 7 msec and
sketch the output waveform. ($V_{cc} = \pm 15$ V)

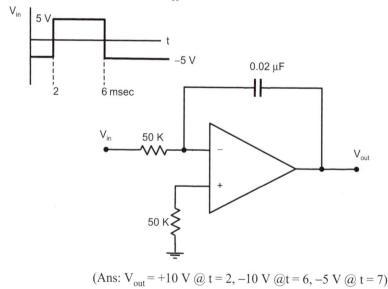

(Ans: $V_{out} = +10$ V @ t = 2, −10 V @t = 6, −5 V @ t = 7)

13. Power supplies

Summary

$$I_{rms} = \frac{H_p l}{\sqrt{2}N}$$ Magnetising current

$$\frac{E_1}{E_2} = \frac{N_1}{N_2}$$ Transformation ratio
$$= a$$

$$P_{max} = V_z \left(\frac{V_{in} - V_z}{R_s} \right)$$ Power dissipation in zener diode

$$\text{Line regulation} = \frac{\delta V_{out}}{\delta V_{in}} \times 100$$

13.1 Transformers

A transformer is a convenient way to step down mains level voltages to those required by electronic circuits.

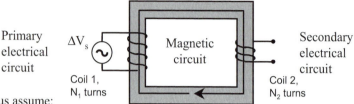

Primary electrical circuit — ΔV_s — Coil 1, N_1 turns

Magnetic circuit

Secondary electrical circuit — Coil 2, N_2 turns

Let us assume:

- the secondary terminals are initially open
- all the flux is confined to the iron core so that flux Φ is same in each coil
- resistance of windings can be neglected

Considering the primary coil:

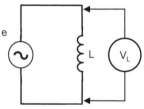

e, L, V_L

Note: using small e to denote instantaneous value of applied emf

Any voltage that appears across the terminals of the coil (effectively an inductor) must be due to the self-induced voltage (**back emf**) in the coil by a changing current through it (**self-inductance**).

$$L\frac{di}{dt} = N_1 \frac{d\phi}{dt}$$

$$= -e_L$$

Instantaneous e_L is proportional to rate of change of <u>current</u> or rate of change of <u>flux</u>

At any instant, $e_L = -e$ by Kirchhoff

thus

$$e = L\frac{di}{dt}$$

$$= V_p \sin(\omega t)$$

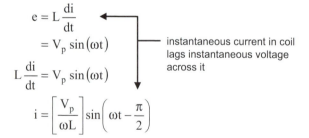

instantaneous current in coil lags instantaneous voltage across it

$$L\frac{di}{dt} = V_p \sin(\omega t)$$

$$i = \left[\frac{V_p}{\omega L}\right] \sin\left(\omega t - \frac{\pi}{2}\right)$$

But the magnetic flux depends directly on the current. Thus, flux in core lags the applied voltage v by $\pi/2$.

13.2 Transformer equations

Now, since the magnetic flux is confined to the core, the same flux passes through each turn of the primary coil. The self-induced voltage in the primary is:

$$e_1 = N_1 \frac{d\Phi}{dt}$$

The flux in the core is sinusoidally varying with time :

Note, it is desirable to express the self-induced voltage in terms of the flux since the flux is the link to the secondary coil (to be examined shortly).

$$\phi = \Phi_p \sin(\omega t - \pi/2)$$

$$\frac{d\phi}{dt} = \omega\Phi_p \cos(\omega t - \pi/2) \quad \text{max when cos term} = 1$$

$$-e_1 = N_1 \omega\Phi_p \cos(\omega t - \pi/2)$$

$$E_{1p} = N_1 \omega\Phi_p$$

This maximum voltage is self-induced in coil #1 by the flux in the core.

$$\boxed{E_{1rms} = \frac{E_{1P}}{\sqrt{2}}} \quad \text{since } V_{rms} = \frac{V_p}{\sqrt{2}}$$

$$E_{1rms} = \frac{N_1 \omega\Phi_P}{\sqrt{2}}$$

For the magnetic circuit in the core:

$$R_m = \frac{f_{m_1}}{\phi} \longrightarrow \text{\textbf{magnetomotive force} (mmf)}$$
established by current in coil 1

$$= \frac{N_1 i_1}{\phi} \quad \text{(lower case denotes instantaneous values)}$$

$$\phi = \frac{N_1 i_1}{R_m} \quad \text{The alternating flux in the core is created by the alternating current in the primary coil.}$$

reluctance of the magnetic circuit

The maximum value of the current is thus:

$$\Phi p = \frac{N_1 I_{1p}}{R_m}$$

$$I_{1p} = \Phi_P \frac{R_m}{N_1}$$

This maximum current induces a maximum magnetic flux in the core.

13.3 Transformer action

For the magnetic flux in the core to be a sine wave, then the flux density in the core must remain in the linear region of the magnetisation curve for the steel being used (i.e. not saturated). The maximum flux will then depend only on the cross-sectional area of the core and the path length.

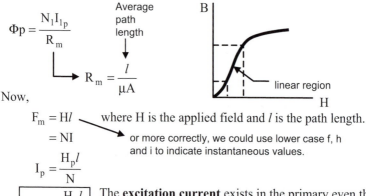

$$\Phi p = \frac{N_1 I_{1p}}{R_m}$$

Average path length

$$R_m = \frac{l}{\mu A}$$

Now,

$$F_m = Hl \quad \text{where H is the applied field and } l \text{ is the path length.}$$

$$= NI \quad \text{or more correctly, we could use lower case f, h and i to indicate instantaneous values.}$$

$$I_p = \frac{H_p l}{N}$$

$$\boxed{I_{rms} = \frac{H_p l}{\sqrt{2}N}}$$ The **excitation current** exists in the primary even though there is no current flowing in the secondary coil. Its function is to establish a changing magnetic flux in the core thus inducing a voltage in the secondary coil.

In the ideal case (no flux leakage), all the flux produced by the primary coil will link with the secondary coil $\Phi_1 = \Phi_2$ and induces a voltage E_2 in the secondary.

$$E_2 = N_2 \frac{d\Phi}{dt}$$

Expressed as an rms voltage, we have:

$$E_{2\,rms} = \frac{N_2 \omega \Phi_p}{\sqrt{2}}$$

> The same changing flux is responsible for the induced voltages in both coils thus the instantaneous and peak voltages e_1, E_{1p} and e_2, E_{2p} are in phase.

The ratio of voltages in the primary and secondary coils is:

$$\frac{E_1}{E_2} = \frac{N_1}{N_2}$$

$$= a$$

Instantaneous, rms or peak values since all in phase

Transformation ratio

Transformers thus provide a way to transform a given AC voltage to another. If a > 1 we have a step-down transformer, if a < 1, a step-up transformer.

13.4 Rectification

The AC output from a transformer is required to be converted to a steady DC value by rectification. This is usually accomplished using diodes in a full-wave **bridge rectifier**.

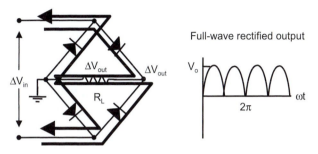

Full-wave rectified output

A rectified output consists of a positive voltage, but varying from 0 V to the amplitude of the transformer secondary output voltage $V_o = E_2$. The output has now to be smoothed to provide a steady DC source for powering electronic circuits.

The easiest and most common method of providing a steady DC source is by the use of a **smoothing capacitor**. The capacitor alternately charges and discharges and serves to smooth out the variations in the output voltage from the rectifier.

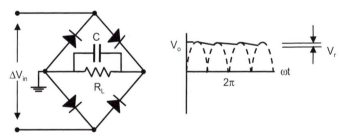

Unfortunately, after smoothing, the output typically contains voltage **ripple** and so must be further conditioned, or regulated to provide an unvarying voltage level.

The peak-to-peak ripple voltage is found from: $V_r = \dfrac{V_o}{2fR_LC}$ f is in Hz

13.5 Regulation

Zener diodes find special application as voltage regulators. They have a very sharp reverse bias breakdown characteristic. In a **voltage regulator**, the supply voltage can change significantly but the zener diode voltage V_Z does not change.

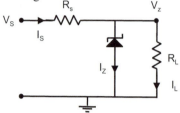

$V_L = V_z$ and I_s is thus fixed and independent of R_L. If R_L increases, the zener passes more current to keep $V_L = V_z$. When R_L is infinite (open circuit), $I_z = I_s$

Maximum current in the zener diode occurs at open circuit conditions where all the current passes through the diode. The **maximum power dissipation** in the diode depends on the input voltage, the breakdown voltage and limiting resistor R_s:

$$P_{max} = V_z I_{max}$$

$$= V_z \left(\frac{V_{in} - V_z}{R_s} \right)$$

If R_L were to decrease, then less current would pass through the zener diode. Since there is a minimum current which must pass through the diode (≈ 5 mA) to maintain operation well into the breakdown region, this limits the amount of current that can be drawn by the load resistor R_L. If the maximum current through the zener is given by I_{max}, and approximately 5 mA is required for reverse breakdown, then the maximum current that can pass through the load resistor R_L is approximately $(I_{max}-5)$ mA. Thus might typically be in the order of 10 mA or so.

A regulator for **high current output** may be constructed using a zener diode in conjunction with an emitter follower circuit.

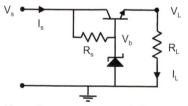

If the load resistance increases, the load voltage V_e increases. An increase in V_e leads to a decrease in V_{be} since the voltage at the base is held constant by the zener diode. The decrease in V_{be} results in an increase in V_{ce} and decrease in V_e. i.e., the output voltage is held constant.

Here. the current output is increased over that normally available by a factor equal to the current gain of the transistor. The diode voltage must be selected to account for the 0.7 V drop across the base-emitter junction.

13.6 Load and line regulation

In voltage regulator circuits such as those shown previously, it is convenient to measure the performance of the circuit in terms of its **load regulation** and **line regulation**.

Load regulation is the percentage change in the output voltage for a change in the load current from zero to full load. In a zener diode circuit, load regulation is affected by the presence of the finite slope of the breakdown region which is called the **dynamic resistance** r_s which might be as much as a few ohms.

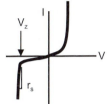

When V_{in} changes, I_Z also changes and so the output voltage is also changed via:
$$V_{out} = V_z + I_z r_s$$

The **line regulation** is the change in the output voltage divided by a change in the input voltage expressed as a percent:

$$\text{Line regulation} = \frac{\delta V_{out}}{\delta V_{in}} \times 100$$

A zener diode regulator incorporating an emitter follower has improved load regulation over that of a simple diode regulator but the line regulation is still limited by the dynamic resistance of the reverse breakdown region.

To overcome the effects of the dynamic resistance of the breakdown region of a zener diode, a second diode may be incorporated into the circuit to act as a pre-stabiliser. In the circuit below, the 10 V zener diode acts as a **pre-stabiliser** for the 4.8 V diode. In this case, the line regulation at the output is considerably improved. This type of circuit acts as a precision **voltage reference** where the output can be used for logic elements or op-amp supplies.

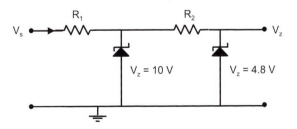

13.7 Thyristor and triac

A **thyristor**, or **silicon controlled rectifier, SCR**, is a voltage regulator with a high breakdown voltage and current gain.

The device has three terminals. Current flows from the anode to the cathode when a pulse is put at the gate terminal. Conduction stops when the supply voltage to the anode is removed whereupon it must be re-triggered at the gate even if voltage is restored to the anode.

This action acts as a rectifier when connected to an AC source. Gate pulses result in current flow until the supply AC falls below 0 V. The timing or phase of the pulses on the gate determine how much of the AC waveform is allowed to pass through the device. When the gate pulses coincide with the positive-going AC half cycle, we obtain a half-wave rectifier action.

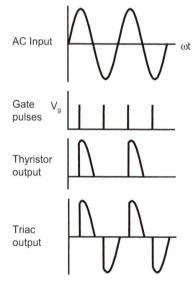

AC Input

Gate pulses V_g

Thyristor output

Triac output

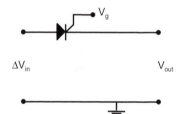

The disadvantage of the half-wave characteristics of the thyristor are overcome by the **triac**. This device consists of two thyristors controlled by one gate. The terminals are referred to as main terminal one MT1 and main terminal two MT2.

Triacs are particularly common in controlling the speed of electric motors or **light dimming**. They also are used in **switched mode power supplies** in which the low frequency mains voltage is rectified directly into a high frequency voltage which is then fed into a step-down transformer and further rectified and regulated. The advantage of operating directly on the mains in this way is that the step down transformer and smoothing capacitors, operating at high frequency, can be made very compact.

14. Laboratory

Experiment 1: Thevenin's theorem

Experiment 2: AC circuits

Experiment 3: Diode characteristics

Experiment 4: Energy gap

Experiment 5: Diode circuits

Experiment 6: Clippers and clamps

Experiment 7: Transistor characteristics

Experiment 8: NiCad Battery charger

Experiment 9: Transistor amplifier

Experiment 10: Amplifier design

Experiment 11: Impedance matching

Experiment 12: FET

Experiment 13: Common source amplifier

Experiment 14: Logic gates

Experiment 15: Logic circuits

Experiment 16: Counters and flip-flops

Experiment 17: Op-amps

Experiment 18: Comparator

Experiment 19: Integrator

Recommended parts required
for laboratory experiments:

Resistors: (× 2 except where noted)
100, 220, 270, 390, 410, 470, 680, 1K, 1.2 K, 1.5K, 2.2K, 2.7K, 3.3K, 3.6K,
4.7K, 6.8K, 10K, 20K, 28K, 36K, 47K, 100K, 270K, 470K, 680K, 1M, 2M
decade resistance box × 1.

Capacitors: (μF, × 2 except where noted)
0.01, 0.82 × 1, 1.0, 4.7, 10, 47, 100 × 1, 1000 × 1

Inductors:
33 mH × 1

Semiconductor devices: (× 1 except where noted)
BC107 BJT × 3
2N5484 FET × 3
4.8 V 1N4732 Zener diode × 3
1N4002 diode × 6
Germanium diode
7400 NAND
7402 NOR
7408 AND
7404 INV
7432 OR
7476 JK flip flop
741 op-amp

Test equipment:
1.5 V dry cell
6 V dry cell
Dual output adjustable power supply ±15 V
Multimeter with mA range and diode test facility × 2
Signal generator 4 kHz
2 channel CRO
Bunsen burner
Beaker of water and tripod
Thermometer

14.1 Thevenin's theorem

Try this simple experiment: Obtain an ordinary AA size 1.5 volt dry cell. Using a voltmeter, measure the voltage between the terminals. Now connect a piece of wire directly from one terminal to the other and measure the voltage between the terminals. Hopefully you will get about 1.5 V for the first measurement and 0 V for the second.

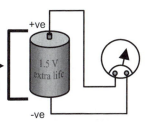

What has happened to the 1.5 V in the second measurement? Since there is no voltage at the output terminals, then all the voltage must be dropped across the **internal resistance** of the cell. If you can, measure the current that flows in the wire when it is connected between the terminals. This is called the "short circuit" current. The voltage measured when there is no wire connected is called the "open circuit" voltage. According to Thevenin, *any* two-terminal output of a power supply circuit containing resistors and voltage sources can be simulated with just a single voltage source V_T and a single resistance R_T. The voltage V_T is usually measured by measuring the open-circuit output voltage V_{oc}. The resistance, R_T, is found by measuring the short-circuit current I_{sc} and calculating: $$R_T = \frac{V_{oc}}{I_{sc}}$$

But, it is not practical to measure short circuit current in most cases since most equivalent resistances are low, making the short circuit currents very high. That is, do NOT try measuring I_{sc} using a car battery.

The short circuit current will be very high and the battery will explode. In practice, measuring short circuit currents could lead to a risk of fire or injury. An alternative procedure is to connect a variable load resistor R_L to the output terminals and adjust it until the output voltage is equal to $V_T/2$. At this condition, $R_T = R_L$

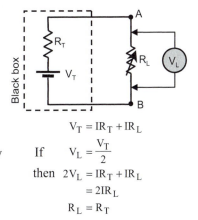

$$V_T = IR_T + IR_L$$
If $$V_L = \frac{V_T}{2}$$
then $$2V_L = IR_T + IR_L$$
$$= 2IR_L$$
$$R_L = R_T$$

Even this procedure is not safe unless the source has a fairly high internal resistance. In the experiment to follow, we shall use a fairly safe 6 V dry cell power supply. Do not attempt measurements of this type unless you are certain that the currents obtained will not cause injury or damage.

Pre-work

Imagine that you are an electrical engineer who has designed the following power supply circuit. When the circuit is used, various "load resistors" are to be connected to the output terminals. Now, to calculate the output terminal voltage and current through the load resistor would be a very tedious task if we had to solve the whole circuit each time we changed R_L, thus, it is easier to find the Thevenin equivalent circuit (V_T and R_T), and then solve a simple current loop for the output voltage and current no matter what value of R_L we might like to consider.

Calculate the numerical value of the Thevenin equivalent circuit ($V_T = V_{ab}$ and R_T) of the circuit below:

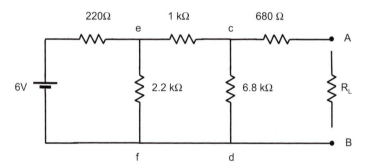

Hint: use a Kirchhoff's law approach by drawing in two current loops and recognising that $V_{AB} = V_{cd}$ in the above diagram at open circuit.

Write your answers here:

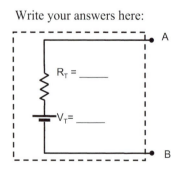

Procedure

1. Construct the circuit shown previously but do not connect R_L (i.e. leave the output at open circuit). Measure the voltage V_{ab} with a voltmeter and record the value.

$$V_T =$$

2. Disconnect the 6 V power supply leads from the circuit. Put a short circuit across the points in the circuit where the power supply used to be connected. Using an ohmmeter, measure and record the resistance between the output terminals AB. Remove the short circuit and re-connect the 6 V power supply.
Compare these measured values with those calculated on the previous page.

$$R_T =$$

3. Using *measured* values for V_T and R_T, use a Thevenin equivalent circuit to calculate the the voltage that will appear across R_L when R_L takes on the values shown :

$R_T =$ _____

$V_T =$ _____

A

B

$R_L = 680\ \Omega$
$R_L = 1.5\ k\Omega$
$R_L = 47\ k\Omega$

V_{AB}

	Calculated	Measured

Connect these resistors in turn to terminals A and B in the actual circuit and measure the voltage V_{ab}. Comment on any differences with calculated values.

4. Connect a decade box resistor as the load resistor in the above circuit. Adjust the decade box so that the voltage V_{AB} is one half of the open circuit voltage. Compare the resistance of the decade box in this condition with the calculated equivalent Thevenin resistance. Make a brief comment about why you observe any correspondence between these two values of resistance.

$$R =$$

14.2 AC Circuits

AC is alternating current, that is, the current in a conductor flows one way, then a very short time later, flows the other way. This transition in direction of flow happens very smoothly. In domestic power lines, the reversal in direction of current is 50 times per second. Here we will look carefully at exactly what AC is and how various components like resistors, capacitors and inductors can resist the flow of alternating current.

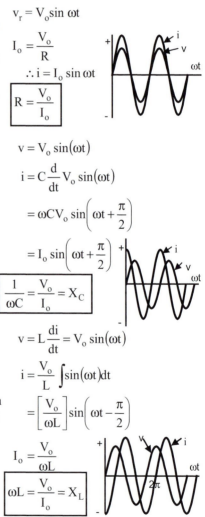

The voltage and current in an AC circuit usually varies sinusoidally. For a resistor, I_o and V_o are in phase and I_o is determined by the value of R just like in a DC circuit.

$$v_r = V_o \sin \omega t$$

$$I_o = \frac{V_o}{R}$$

$$\therefore i = I_o \sin \omega t$$

$$R = \frac{V_o}{I_o}$$

For a **capacitor**, the resistance to AC, and thus the magnitude of I_o for a given V_o, depends on the value of the capacitor *and* the frequency of the applied voltage. At high frequencies, the polarity of the voltage applied to the capacitor changes before the capacitor has had a chance to charge up, thus the current I_o is large. Thus, the reactance of a capacitor decreases with increasing frequency. Further, the maximum voltage occurs when the current is zero and decreasing , thus I_o leads V_o by $+\pi/2$.

$$v = V_o \sin(\omega t)$$

$$i = C \frac{d}{dt} V_o \sin(\omega t)$$

$$= \omega C V_o \sin\left(\omega t + \frac{\pi}{2}\right)$$

$$= I_o \sin\left(\omega t + \frac{\pi}{2}\right)$$

$$\frac{1}{\omega C} = \frac{V_o}{I_o} = X_C$$

For an **inductor**, the magnitude of I_o again depends on the frequency of V_o. For high frequencies, the magnitude of the induced back emf is large and this restricts the maximum current that can flow before the polarity of the voltage changes over. Thus, the reactance increases with increasing frequency. The maximum voltage occurs when the rate of change of current is a maximum and increasing, thus I_o lags V_o by $\pi/2$.

$$v = L \frac{di}{dt} = V_o \sin(\omega t)$$

$$i = \frac{V_o}{L} \int \sin(\omega t) dt$$

$$= \left[\frac{V_o}{\omega L}\right] \sin\left(\omega t - \frac{\pi}{2}\right)$$

$$I_o = \frac{V_o}{\omega L}$$

$$\omega L = \frac{V_o}{I_o} = X_L$$

Procedure: Part A. Reactance

1. Construct the circuit shown. Set the function generator to 1V peak to peak at sine wave of 500 Hz.

2. Set the multimeters to AC operation and measure the rms voltage across the 270 Ω resistor and also the rms current. Now adjust the amplitude of the sine wave to obtain 0.5 V rms on the voltmeter.

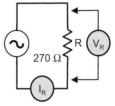

3. Record values of rms voltage and current for each of the frequencies as shown in the table. **Keep the rms voltage at 0.5 V by adjusting the amplitude each time you change the frequency.**

4. Replace the resistor with a 1 μF capacitor and measure rms voltage and current as a function of frequency. Keep rms voltage at 0.5 V at each frequency step. Record results in the table.

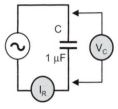

5. Replace the capacitor with the inductor supplied and measure the rms voltage and current as a function of frequency as above. Record results in the table.

6. *Calculate* the resistance or reactance for each device at each frequency using the experimentally measured rms voltages and currents $X_C = V_{rms}/I_{rms}$.

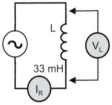

7. Verify for one set of readings (say 500 Hz), that the value for reactance calculated using rms values corresponds to that using the formula on the previous page. i.e. $X_C = 1/\omega C$, $X_L = \omega L$.

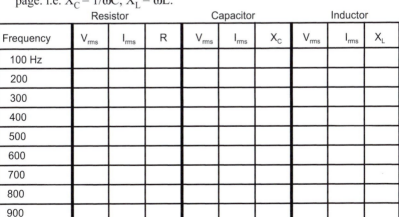

Frequency	Resistor			Capacitor			Inductor		
	V_{rms}	I_{rms}	R	V_{rms}	I_{rms}	X_C	V_{rms}	I_{rms}	X_L
100 Hz									
200									
300									
400									
500									
600									
700									
800									
900									

Procedure: Part B. High pass filter

1. Construct the circuit below.

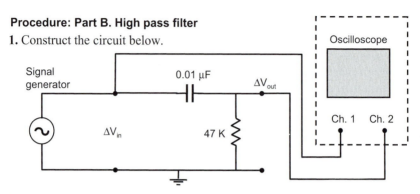

2. Set the signal generator output for ΔV_{in} to be a sine wave, 1 V peak-to-peak, at 50 Hz. If you have wired up the circuit correctly, the input signal will appear on Channel 1 of the CRO and the output signal on Channel 2.

3. Once you have got the CRO settings adjusted so that you can view both input and output signals, measure and record the peak-to-peak output voltage as a function of frequency (50, 100, 200, 300, 400, 500, 1000, 2000, 4000 Hz). **Make sure that the input signal ΔV_{in} is maintained at 1 V peak-to-peak. Adjust the amplitude of the signal generator if necessary.** Enter measured values of V_{out} in table below.

$\Delta V_{in} = $ _____

All ΔV values are peak-to-peak.

Freq. (Hz)	ΔV_{out} (Measured)	ΔV_{out} (Calc)
50		
100		
200		
300		
400		
500		
1000		
2000		
4000		

Calculations:

$$\frac{\Delta V_{out}}{\Delta V_{in}} = \frac{R\omega C}{\sqrt{R^2\omega^2 C^2 + 1}}$$

Let $R\omega C = 1$

$$\frac{V_{out}}{V_{in}} = \frac{1}{\sqrt{2}}$$

3 dB point (by definition)

4. Using the nominal values of the components you are using, calculate values of output voltage for each of the frequencies used in the measurements with an input voltage of 1V peak-to-peak.

5. Calculate the 3 dB frequency for this circuit. Enter data in table above and 3 dB frequency in box on next page.
Hint: examine calculations shown above to find ΔV_{out} as a function of ΔV_{in}, R, C and ω and thus calculate ΔV_{out} for each frequency step.

6. Plot a graph of $\Delta V_{out}/\Delta V_{in}$ as a function of frequency for both measured
and calculated values (plot on same graph) using log/linear graph paper.
Determine the "measured" 3 dB point from the appropriate graph and
enter in the table below:

Frequency @
3 dB point:

Measured:

Calculated:

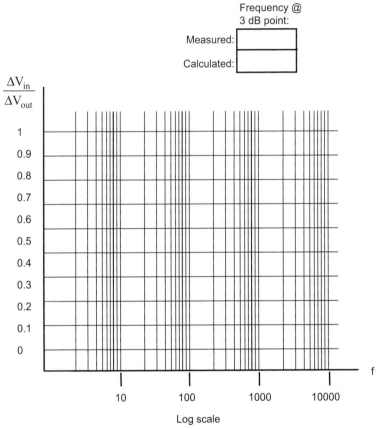

$$\frac{\Delta V_{in}}{\Delta V_{out}}$$

1
0.9
0.8
0.7
0.6
0.5
0.4
0.3
0.2
0.1
0

f

10   100   1000   10000

Log scale

7. Comment on any features of interest in your graphs and on any
differences you see between measured and expected values.

14.3 Diode characteristics

Diodes were the first semiconductor electronic devices. A **crystal** in a crystal set radio is a diode. Diodes are used to convert AC signals to DC signals, protect electronic equipment from voltage fluctuations, provide stable DC voltage levels and many other functions. How can such a simple device accomplish so much?

A p-n junction will conduct current in forward bias and act as an insulator in reverse bias. Such a process is called **rectification** and the device as a whole is called a **diode**.

A perfect diode would present zero resistance in forward bias and infinite resistance in reverse bias.

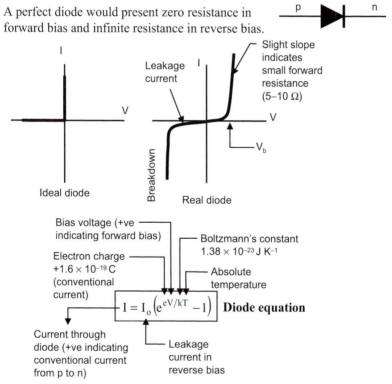

There are different types of diodes. A **zener diode** is specially constructed so that it has a well-defined reverse bias breakdown voltage. The value of the breakdown voltage can be set during manufacture (where the forward bias voltage is fixed by the barrier potential of the semiconductor material). Zener diodes are usually connected in reverse bias and can be used to regulate voltage.

Procedure: Part A Normal diode

1. Construct the circuit shown below using a normal diode. Use the bi-polar adjustable supply from the laboratory power supply.

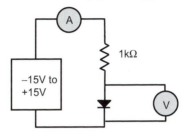

2. Adjust the power supply so that the diode is placed in reverse bias with a voltage of –6 V. Set the ammeter on the most sensitive scale (200 μA)

3. Measure and record the current through the diode for voltages from –6 V to 0 V in 2 V steps.
   Be careful not to burn out the fuse on the ammeter - adjust the voltage slowly.

V	I μA
-6	
-4	
-2	
0	

4. Change to a higher scale on the ammeter, say 200 mA. SLOWLY increase the voltage in 0.1 V steps up to a value of +0.5 V and then in 0.05 V steps to 0.8 V **or** a maximum of 15 mA.

WARNING: Do not run a large current (> 15 mA) through the diode. It will burn out.

V	I
+0.1	
+0.2	
+0.3	
+0.4	

V	I
+0.5	
+0.55	
+0.6	
+0.65	

V	I
+0.7	
+0.75	
+0.8	

5. Plot a graph of current (μA) versus voltage (−6 to 0).

6. Plot a graph of current (mA) versus voltage (0 to +0.8).

7. Determine as accurately as possible the leakage current in reverse bias.

8. Using the diode equation, plot on the same graph as above the expected characteristics of the diode.

9. Comment on any other features you think are significant.

Part B. Zener diode

1. Determine the I-V characteristic for a zener diode using the voltage steps as suggested below.

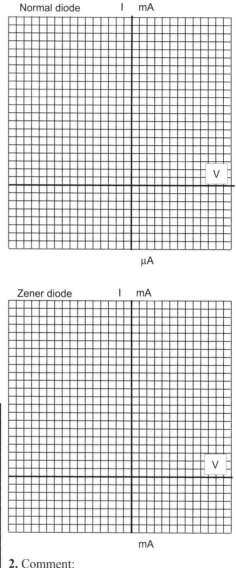

Forward bias		Reverse bias	
V	I mA	V	I mA
0		-0.5	
0.1		-1.0	
0.2		-1.5	
0.3		-2	
0.4		-2.4	
0.5		-2.6	
0.6		-2.8	
0.7		-3	
0.75		-3.2	
		-3.4	
		-3.6	
		-3.8	

2. Comment:

14.4 Energy gap

In this experiment, you will determine the energy gap in germanium and silicon by investigating the temperature dependence of the electrical conductivity of a forward biased p-n junction.

Maxwell-Boltzmann statistics applied to diffusion of charge carriers can predict current density across junction in forward and reverse bias. Analysis of this type shows that the current through a diode in forward bias can be described mathematically by the **diode equation**.

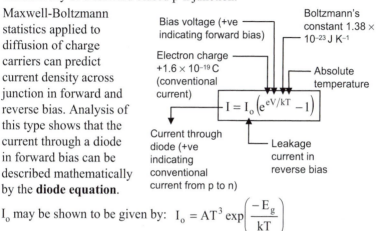

I_o may be shown to be given by: $I_o = AT^3 \exp\left(\dfrac{-E_g}{kT}\right)$

where A is a constant and E_g is the energy gap of the material.

Combining this with the diode equation gives:

$$V = \frac{E_g}{e} + \frac{k}{e}\ln\left(\frac{I}{A}\right)T - \frac{3kT}{e}\ln T$$

which may be re-arranged to give the following linear form:

$$V + \frac{3kT}{e}\ln T = E_g' + \frac{k}{e}\ln\left(\frac{I}{A}\right)T$$

where the energy gap is expressed in electronvolts

Procedure:

1. Connect the germanium diode to the digital voltmeter and turn the voltmeter to the diode setting. This setting is indicated on the meter by the diode symbol. (See note below).

2. Place the germanium diode adjacent to a thermometer bulb. Place both the diode and the thermometer into a beaker of water. Make measurements of the temperature of water and the voltage across the diode between 0 °C and 100 °C. Stirring the water will help to assure uniform temperatures throughout.

3. Repeat the above procedure using a silicon diode.

4. Plot a graph of $V + \dfrac{3kT}{e} \ln T$ on the y axis against T (in K) on the x axis. You should obtain a straight line whose intercept is the energy gap E_g (in eV).

5. Answer the following questions:

    (a) What are the accepted values for E_g for germanium and silicon? Calculate the percentage difference between the values of E_g you obtained and the accepted values.

    (b) Show clearly the algebraic steps necessary to obtain the linearised form of the equation given in the box on the previous page.

    (c) From the gradient of the graphs find A for the germanium and silicon diodes.

    (d) Is E_g temperature dependent?

    (e) What are the major sources of error involved in the experiment? and how may these errors be minimised?

Caution: The diodes should be **forward biased** during this experiment. This can be done by making sure that the black plug connected to the diode is inserted into the 'common' (or com) socket of the meter and the red plug is inserted into the volts socket of the meter. When the meter is on the diode setting, the voltage appearing on the meter should be between 0.2 and 0.5 V.

14.5 Diode circuits

Your portable radio/CD player operates off batteries, right? But,you can also run it off the power point at home to save batteries if you wish. How can this be? Batteries provide DC but the power point is 240 V AC. The answer of course is that the AC is converted into DC either inside the machine, or in an external **battery saver** which uses diodes.

The conversion of AC voltage to DC voltage is called **rectification**. Most portable equipment uses low voltage DC. A transformer may be used to produce low voltage AC from the 240 V AC mains, but this then has to be rectified to obtain a steady DC output. Simple rectification can be had with just a single diode.

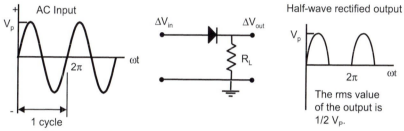

Although we have obtained a positive going output, it is by no means a very steady one. A better output can be obtained with a "bridge" rectifier and offers "full-wave" rectification. Full-wave rectification involves a clever arrangement of 4 diodes to produce a DC signal but with a large ripple.

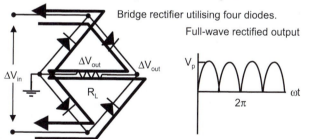

Although we still have a fairly non-steady output, it is an vast improvement on the half-wave design and may be easily modified to give a smooth DC output voltage.

Rectification involves the use of diodes connected in forward bias. But, you say, what about **zener diodes**? They break down at fairly low reverse bias voltages, what can they be used for? Zeners are used for voltage regulation, and we will look at this application of diodes later in this experiment.

Procedure: Part A. Voltage rectification

1. Construct the circuit shown below using normal diodes. Adjust the signal generator to give a 5 volt peak-to-peak sine wave output at a "reasonable" frequency. Attach the oscilloscope probes so that the input and output signals can be displayed on the monitor simultaneously.

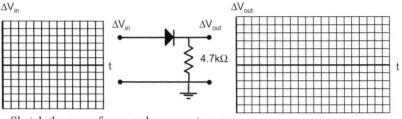

Sketch the wave forms and comment on any features of interest.

2. Now construct a full-wave rectifier circuit using the 4.7 kΩ load resistor. Apply the the 5 V peak-to-peak signal from the signal generator and sketch the wave forms.

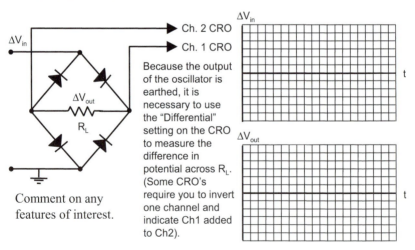

Because the output of the oscillator is earthed, it is necessary to use the "Differential" setting on the CRO to measure the difference in potential across R_L. (Some CRO's require you to invert one channel and indicate Ch1 added to Ch2).

Comment on any features of interest.

3. Modify the circuit by adding components where you think necessary to obtain a steady DC voltage across R_L. Describe what you think is happening and the significance of the value of the component(s) you used to smooth this output.

Procedure: Part B. Voltage regulation (zener diode)

Zener diodes find special application as voltage regulators. They have a very sharp reverse bias breakdown characteristic. In a voltage regulator, the supply voltage can change significantly but the zener diode voltage V_Z does not change.

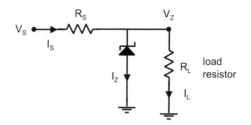

$V_L = V_Z$ and I_S is thus fixed and independent of R_L. If R_L increases, zener passes more current to keep $V_L = V_Z$. When R_L is infinite, $I_Z = I_S$.

But, useful as they are, the diodes are only capable of passing a certain maximum amount of current before they overheat. Thus, we need to have a current limiting resistor in series with the device to limit the maximum current (and hence power dissipation) in the device.

1. For the circuit shown below, calculate the value of a current limiting resistor R_Z required so that the power dissipated by the 4.8 V zener diode does not exceed 25 mW.

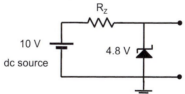

2. Construct the circuit and attach a decade resistance box as a variable load resistor R_L.

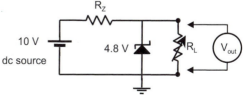

3. Starting from 10 kΩ, decrease the value of R_L in appropriate steps until the voltage V_{out} drops to about 70% of its original value. Record your readings in the table.

Resistance	V_{out}	Resistance	V_{out}

4. Plot this data and <u>comment on the significance of your findings</u>.

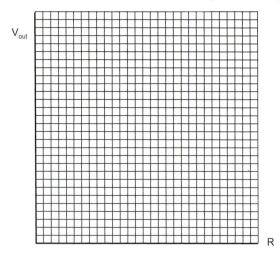

V_{out}

R

14.6 Clipper and clamps

Ever heard a **Geiger counter** click? Ever wondered how your computer can obtain a stable voltage for the internal circuitry from the mains supply which varies according to what appliances are operating nearby? These are examples of simple diode circuits in action.

In order to protect sensitive circuitry from high voltages, it is usual to incorporate diode **clippers** into circuits. In this lab we will build a clipper and measure its characteristics. We will also build another very useful circuit, the diode **clamp**, in which a DC level shift can be applied to a signal.

Pre-work

Inspect the following circuits 1 to 5 and calculate the expected output voltage of each.

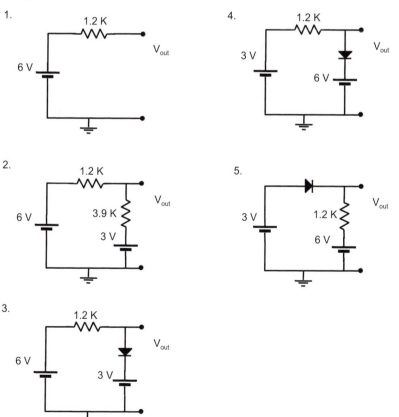

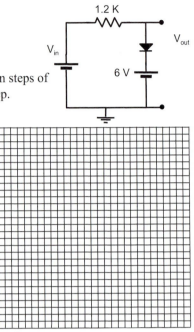

Procedure: Part A. Clipper

1. Construct the circuit shown.
2. Increase V_{in} from −14 V to +14 V in steps of
   2 volts and measure V_{out} at each step.
2. Plot a graph of V_{out} versus V_{in}.

V_{in}	V_{out}	V_{in}	V_{out}

3. How would you modify the circuit so that V_{out} clipped at 2 V ?
4. Do that modification and check whether clipping does occur at 2 V.
5. Replace the DC power supply labelled V_{in} in this circuit with an oscillator.
   Set to sinusoidal output with ΔV_{in} at 10 V peak to peak (use oscilloscope
   to measure ΔV_{in}). Sketch the output waveform.
6. Repeat step 5 with ΔV_{in} at 20 V peak to peak. Sketch output waveform.

Procedure: Part B. Clamp

1. Construct the circuit shown.
2. Set V_{in} to 10 V peak to peak and show
   V_{in} and V_{out} on an oscilloscope.
2. Sketch the waveform that results.
3. Explain any differences you observe
   between V_{in} and V_{out}

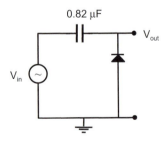

14.7 Transistor characteristics

The transfer resistor, or **transistor**, was invented in 1947 by scientists working at the Bell Telephone Laboratories in the United States. It revolutionised the field of electronics since it allowed amplification of electrical signals to take place using a small, low power, robust device which eliminated the need for bulky, fragile vacuum tubes. In this experiment, we examine the electrical characteristics of a common npn silicon transistor which can be purchased for about 50c.

How does it work?

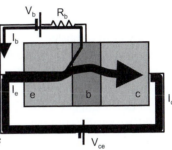

1. Base-emitter junction is a forward-biased p-n junction so when $V_{be} > 0.7$ V, then junction becomes conducting (just like a diode).

2. Electrons coming from the heavily doped emitter cross junction but before they have a chance to combine with holes in the p-type base and travel to the V_b positive terminal, they get swept up by the strong field which exists around the collector base junction which is reverse-biased.

3. Hence, only a few electrons go towards +ve V_b and most are attracted across the collector base junction and cause a large current in the collector.

The base is made lightly doped (so that recombination in the base is unlikely to occur) and is purposely made very thin (so that electrons coming across the forward-biased b-e junction do not have far to go before they "overshoot" and fall into the field across the c-b junction).

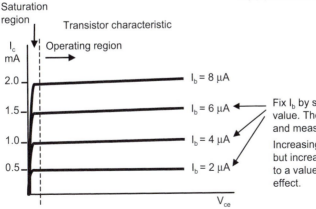

Saturation region

Transistor characteristic

I_c mA

Operating region

2.0 — $I_b = 8 \ \mu A$

1.5 — $I_b = 6 \ \mu A$ ← Fix I_b by setting V_b to some value. Then increase V_{ce} and measure I_c.

1.0 — $I_b = 4 \ \mu A$

0.5 — $I_b = 2 \ \mu A$

Increasing I_b increases I_c but increasing V_{ce} (once set to a value > 0.1 V) has no effect.

V_{ce}

Procedure:

1. Construct the circuit shown using two independent power supplies so that V_{ce} and I_b can be controlled independently. Adjust I_b to 2 μA and record values for I_c at V_{ce} = 15 V, 10 V, 5 V, 2 V, 1 V, 0.5 V and 0.1 V.

2. Repeat these measurements for I_b = 4 μA, 6 μA, 8 μA and 10 μA and record results in a table.

3. Construct a plot (below) showing the transfer characteristic at V_{ce} = 5V and determine the current gain (h_{fe}).

4. Draw a graph of I_c vs V_{ce} for different values of I_b and identify the operating and saturation regions (next page).

Note that $I_e = I_b + I_c$ but since $I_b \ll I_c$, then $I_e = I_c$

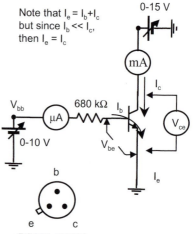

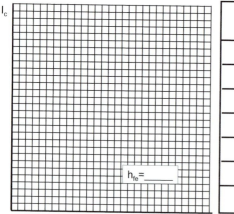

BC107, BC108

h_{fe}=_____

V_{ce}	I_b = 2	4	6	8	10 μA
15					
10					
5					
2					
1					
0.5					
0.1					

I_c

I_b

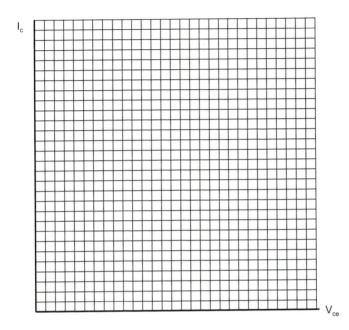

5. Comment on any features of the above transistor characteristic that you think are significant.

6. Examine the effect of reducing V_{ce} down to zero volts on the base current. That is, what value of I_b do you obtain when $V_{ce} = 0$. Can you explain your reading?

14.8 NiCad battery charger

Rechargeable batteries are widely used in portable equipment. The most popular type are NiCad cells. The best operating conditions for NiCad batteries are regular use. However, NiCad batteries, unlike lead-acid batteries, can only be charged in a "short" time (usually about 15 hours) using a **constant current**. After the specified charge time, the charging current must be turned off. The constant current characteristics of a transistor can be used to make a NiCad battery charger.

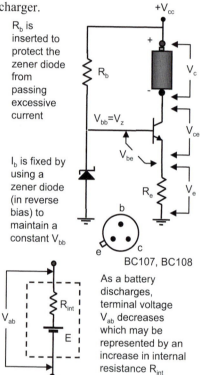

How does it work? Consider the circuit shown at the right:

R_b is inserted to protect the zener diode from passing excessive current

1. The battery to be charged is placed in the circuit in lieu of R_c.

2. The charging current is set by the value of the zener diode to some "design value", say 10 mA.

3. When a "flat" battery is first inserted, its internal resistance is high (same as a high R_c) and thus the "charging" voltage V_c is high and V_{ce} is low to maintain constant I_c.

   I_b is fixed by using a zener diode (in reverse bias) to maintain a constant V_{bb}

4. As battery terminal voltage increases (i.e. decreasing "R_c") charging voltage decreases and V_{ce} increases.

5. When battery terminal voltage reaches a maximum i.e. fully charged, (R_{int} levels off to some "low" value) V_c is low and V_{ce} is high.

   As a battery discharges, terminal voltage V_{ab} decreases which may be represented by an increase in internal resistance R_{int}

Throughout this process, and even when battery is fully charged, I_c is still flowing so another circuit must switch off after a specified time period. However, having I_c a constant during charging ensures that battery is charged at the maximum rate and to full capacity. Some NiCad battery chargers do not use a constant current source but supply a "trickle" charge. Trickle chargers can be left on for days without harm to the cells, but the cells would take 2 or 3 days to reach full charge from a completely discharged condition.

Procedure:

1. Calculate the values of R_e and R_b required for a constant current of $I_c = 10$ mA using a supply voltage of 10 V and a 4.8 V zener diode. The Zener diode is to pass a current of no more than 20 mA. Show all calculations.

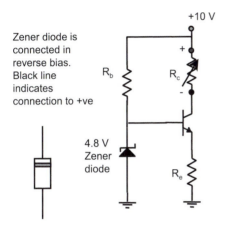

Zener diode is connected in reverse bias. Black line indicates connection to +ve

+10 V

R_b R_c

4.8 V Zener diode

R_e

Hint: I_b is very much smaller than the current through the zener.

2. Construct the circuit and measure the current I_c with the "load" resistance R_c set to 0 Ω. Make sure the ammeter is connected between V_{cc} and the collector. Do not connect the ammeter between the emitter and R_e or between R_e to ground. I_c should be approximately 10 mA. Seek help if this is not the case.

Measured value I_c at R_c=0:

3. Using the decade box as a load resistor, record the current I_c starting from about 10 kΩ (or higher if possible) decreasing to 0 Ω in steps of about 500 Ω initially but decreasing the step size near where the current I_c begins to stabilise. Measure the value of V_{ce} at which the charging current I_c becomes fairly constant (should be approx 10 mA).

R_c	I_c	R_c	I_c	R_c	I_c

4. Write down an equation for the load line for this circuit.
5. Determine the x and y axes intercepts for selected load lines from your data (i.e. different values of R_c) and indicate on a graph of I_c vs V_{ce} these load lines. Make sure you include some load lines which correspond to *the beginning* of a fall-off in collector current.
6. Comment on why there is this fall-off in I_c and the usefulness of the circuit as a constant current source.

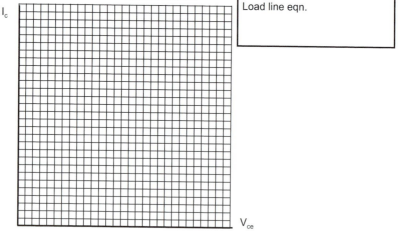

Load line eqn.

7. In the analysis of this circuit, it was assumed that the base current was always much less than the collector current. However, this is an unusual circuit due to the inclusion of the resistor at the emitter together with the zener diode at the base. These two conditions impose a "feedback" effect on the circuit, the effect of which may be demonstrated by removing R_c from the circuit (or making it a very high value) and measuring either the base or the emitter current. Measure the emitter current at high value of R_c (or simply remove it from the circuit and leave the collector at open circuit).

Your observation:

8. In the light of the above observation, it is evident that the base current does play an important role at very low values of I_c. With this in mind, examine the circuit again, and this time, develop a more accurate expression for the load line which includes the base current in the expression.

9. What does this mean about the slope and intercepts of the load line for this circuit when the transistor is saturated?

14.9 Transistor amplifier

The transistor can be used as a switch, a source of constant current, and to amplify small signals. Ever tried connecting your headphones directly to the "line out" of your tuner or CD deck? You won't get much volume. The signal has to be amplified before it is large enough to drive a loudspeaker and this is where a transistor comes into its own.

Here is an amplifier circuit you have been studying in this book. Using the step-by-step approach, calculate the DC bias conditions and AC performance of the circuit.

(Note, you will need to measure or estimate a value of h_{fe} for the transistor you intend to use to build this circuit).

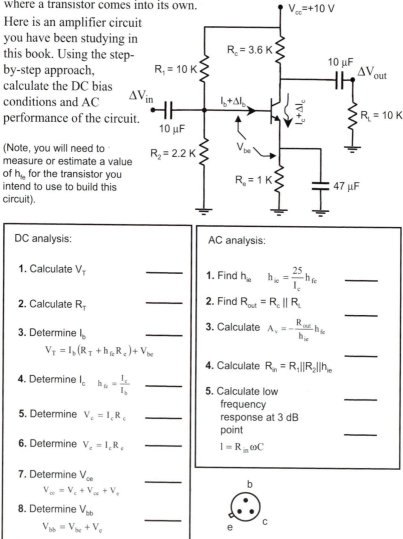

DC analysis:

1. Calculate V_T _____

2. Calculate R_T _____

3. Determine I_b _____

$$V_T = I_b(R_T + h_{fe}R_e) + V_{be}$$

4. Determine I_c $\quad h_{fe} = \dfrac{I_c}{I_b}$ _____

5. Determine $V_c = I_c R_c$ _____

6. Determine $V_e = I_c R_e$ _____

7. Determine V_{ce} _____

$$V_{cc} = V_c + V_{ce} + V_e$$

8. Determine V_{bb} _____

$$V_{bb} = V_{be} + V_e$$

AC analysis:

1. Find h_{ie} $\quad h_{ie} = \dfrac{25}{I_c}h_{fe}$ _____

2. Find $R_{out} = R_c \| R_L$ _____

3. Calculate $A_v = -\dfrac{R_{out}}{h_{ie}}h_{fe}$ _____

4. Calculate $R_{in} = R_1 \| R_2 \| h_{ie}$ _____

5. Calculate low frequency response at 3 dB point _____

$$1 = R_{in}\,\omega C$$

Procedure:

1. Construct the circuit using a 10 kΩ resistor as R_L and measure the DC bias conditions (V_c, V_{ce}, V_e, V_{bb}, I_c and I_b) and AC voltage gain A_v and enter measured values in table. Summarise your readings and calculations in the table below and thus comment on any discrepancies you encounter.

	Calculated	Measured
V_{cc}		
I_b		
I_c		
V_c		
V_e		
V_{ce}		
V_{bb}		
A_v		

Note: do not measure the AC voltage gain at the 3 dB point frequency. Select a mid-range frequency and measure ΔV_{in} and ΔV_{out} You decide a suitable input voltage ΔV_{in}

ΔV_{in} used:
Frequency used:

Comments & results:

ΔV_{in}

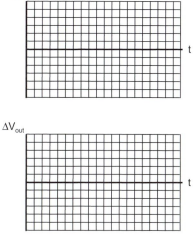

t

ΔV_{out}

t

2. Measure the frequency response of your circuit and check the 3 dB point. Do this by connecting channel 1 on the CRO to the input and channel 2 to the output. Decrease the frequency and note the frequency at which the signal on channel 2 becomes approx 70% of that of the mid-frequency response.

Frequency @ 3 dB point: ☐

3. Remove the <u>bypass</u> capacitor and measure the AC voltage gain at a mid-range frequency. Is this consistent with what you would calculate with no bypass capacitor?

With no bypass capacitor:

$A_{v\ calculated}$: ☐ $A_{v\ measured}$: ☐

Comment on the effect of the bypass capacitor.

14.10 Amplifier design

In the last experiment, you were given a circuit to build and analyse. How did the designer of the circuit know what values of resistances to use? How was the voltage gain decided? Here's your chance to design your own amplifier.

Firstly, you have to do some calculations. Use the step-by-step approach to design a common emitter amplifier with the following characteristics:

• open circuit voltage gain = 200 (with no load)
• input resistance (h_{ie}) 3 kΩ
• lower frequency limit 50 Hz

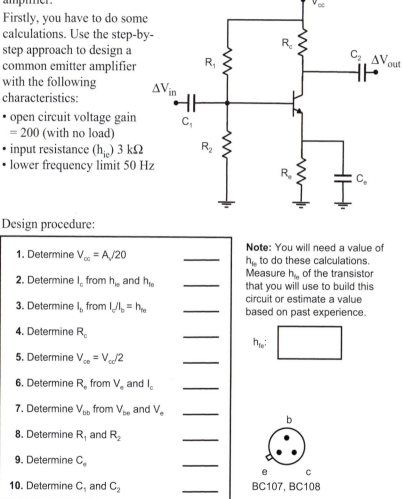

Design procedure:

1. Determine $V_{cc} = A_v/20$ _____

2. Determine I_c from h_{ie} and h_{fe} _____

3. Determine I_b from $I_c/I_b = h_{fe}$ _____

4. Determine R_c _____

5. Determine $V_{ce} = V_{cc}/2$ _____

6. Determine R_e from V_e and I_c _____

7. Determine V_{bb} from V_{be} and V_e _____

8. Determine R_1 and R_2 _____

9. Determine C_e _____

10. Determine C_1 and C_2 _____

Note: You will need a value of h_{fe} to do these calculations. Measure h_{fe} of the transistor that you will use to build this circuit or estimate a value based on past experience.

h_{fe}:

BC107, BC108

Procedure:

1. Construct the circuit and measure DC bias conditions, AC voltage gain
and low frequency cut-off.

Resistors

	Calculated	Measured
V_{cc}		
I_b		
I_c		
V_c		
V_{ce}		
V_e		
V_{bb}		
V_{be}		
A_v		

Resistors	Calculated	Used
R_c		
R_e		
R_1		
R_2		

Capacitors	Calculated	Used
$C_{1,2}$		
C_e		

2. Examine the effects of increasing the input signal ΔV_{in}. What is the
maximum peak-to-peak input signal that may be tolerated before
clipping appears on the output? Explain why this clipping occurs.

3. Examine the effect of heating the transistor on h_{fe}. Insert the transistor into
the h_{fe} measurement instrument and measure h_{fe}. Then note any change in
h_{fe} when touching (firmly) the case of the transistor. Is there any effect?
Now insert the transistor into the amplifier circuit and determine whether
there is any effect on the voltage gain of the circuit when the transistor is
heated (use fingers again). Explain what you observe.

14.11 Impedance matching

In previous experiments, you found that the voltage gain of a common emitter amplifier depended upon whether or not a "load" resistor (which might be say a loudspeaker) is connected at the output. To some extent, the voltage gain also depends on the internal resistance of the signal source. How can these effects be calculated and optimised?

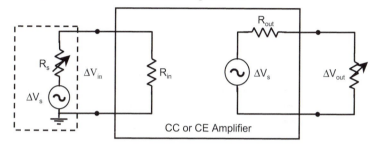

The circuit shown above represents the equivalent circuit of an amplifier. The open circuit gain is found from:

$$A_v = \frac{\Delta V_o}{\Delta V_i}$$

To measure the input and output resistances of the amplifier, the following approach is adopted:

BC107, BC108

1. Input resistance R_{in}.

With the output on open circuit (R_L = infinity) and the variable resistance $R_s = 0$, measure the input voltage ΔV_{in}. Gradually increase the variable resistance R_s until the value of ΔV_{in} is halved. At this point, $R_s = R_{in}$.

2. Output resistance R_{out}

With the output on open circuit (R_L = infinity) record the value of ΔV_{out} and then decrease the value of R_L until the output voltage ΔV_{out} is halved. At this point, $R_L = R_{out}$.

In this experiment, you will compare the input and output resistance and voltage gain of both common emitter and common collector circuits. The input and output resistances will be determined using the procedures 1 and 2 above for both circuits in turn. The calculated or expected values may be found from:

For **common emitter**:	For **common collector**:
$h_{ie} = \dfrac{h_{fe}\,25}{I_c}$	$R_{in} = h_{ie} + h_{fe}R_e$
$R_{out} = R_c \| R_L$	$R_{out} = \dfrac{R_s + h_{ie}}{h_{fe}}$

Procedure: Part A. Common emitter circuit

1. Consider the circuit shown below. Assuming that $V_{be} = 0.7$ and that V_{ce} is to be $V_{cc}/2$, determine an expression for R_b as a function of h_{fe}. Measure h_{fe} for the transistor that you will use in this laboratory session and thus calculate a suitable value of R_b to use in this experiment.

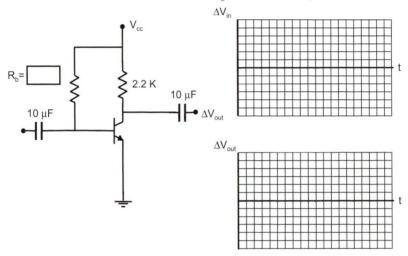

2. Construct the circuit and measure the open circuit gain at 5 kHz using an input signal in the vicinity of 30 mV pk-to-pk with $V_{cc} = +10$ V. Make sure that there is no saturation on the output (clipping). A_v [   ]

3. Insert a variable resistor R_s between the signal generator and the amplifier input. Measure and record the input resistance using the method described on the previous page. R_{in} [   ]

4. Disconnect the variable resistor from the input and connect it to the output terminal. Measure the output resistance of the amplifier using the method described on the previous page. R_{out} [   ]

5. Calculate the expected R_{in} and R_{out} and compare with experimental readings.   Expected values:

R_{in} [   ] R_{out} [   ]

Procedure: Part B. Common collector circuit

1. Using the same components as above, construct a common collector configuration as shown:

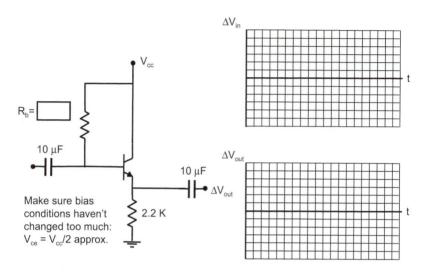

2. Using a 30 mV pk-to-pk input signal, measure the AC gain at $V_{cc} = +10$

A_v

3. Measure the input and output resistances of this circuit using the same method as used for the CE amplifier circuit.

Measured values:

R_{in}

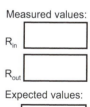

4. Determine the expected values and compare with measured values.

R_{out}

Expected values:

R_{in}

R_{out}

5. Make a comment on the significance of R_{in} and R_{out} for the two circuits you have examined in this experiment.

14.12 FET

So far we have concentrated on the use of a **bipolar junction transistor** - a current controlled device in which the collector current is controlled by the base current. Now we will have a look at **field effect transistor** where the current is controlled by a voltage signal rather than a current. This leads to certain advantages when one requires a high impedance input.

How does it work?

As V_{gs} is made more negative, the channel narrows as the depletion layer widens and this constriction reduces the drain current I_d.

For a given V_{ds}, the drain current I_d depends on V_{gs}. At $V_{gs} = 0$, I_d is a max (no depletion region - channel is widest). The correspondence between V_{gs} and I_d is described by the transfer characteristic.

The relationship between I_d and V_{ds} is shown on the transistor characteristic. For FET's, there is a noticeable slope in the normal operating region. This corresponds to a "resistance" within the conducting channel and may need to be taken into consideration when determining the voltage gain of a circuit employing an FET.

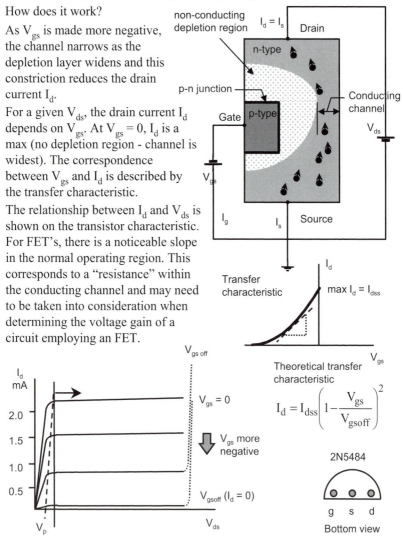

non-conducting depletion region $I_d = I_s$ Drain

n-type

p-n junction

Gate   p-type

Conducting channel

V_{ds}

V_{gs}

I_g I_s Source

Transfer characteristic

I_d

max $I_d = I_{dss}$

$V_{gs\ off}$

$V_{gs} = 0$

V_{gs} more negative

V_{gsoff} ($I_d = 0$)

I_d mA

2.0

1.5

1.0

0.5

V_p

V_{ds}

Theoretical transfer characteristic

$$I_d = I_{dss}\left(1 - \frac{V_{gs}}{V_{gsoff}}\right)^2$$

2N5484

g   s   d

Bottom view

Procedure: Part A. Transfer characteristic

1. Construct the circuit shown below **(read the warning message below before applying power to circuit).**
2. Measure the drain current I_d in mA as a function of the gate-source voltage V_{gs} from $V_{gs} = 0$ V to -1.4 V in steps of -0.1 V (or until $I_d < 0.1$ mA).
3. Plot I_d against V_{gs} and identify the range in V_{gs} over which I_d is approximately linear.
5. Determine the mutual transconductance g_m in mS in the linear region of the characteristic
6. Estimate values of V_{gsoff} and I_{dss} and plot the theoretical transfer characteristic.

> **WARNING!** It is very easy to burn out the JFET in this experiment. DO NOT MAKE V_{gs} +ve.

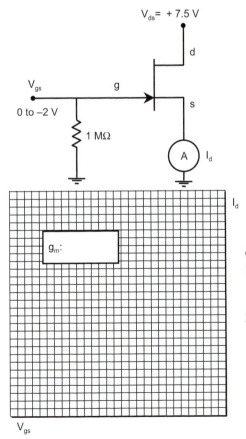

$V_{ds} = +7.5$ V

V_{gs}
0 to -2 V

$1\ M\Omega$

g

d

s

A I_d

g_m:

V_{gs}

I_d

V_{gs}	I_d	V_{gs}	I_d

Questions:
1. Why is the drain current a maximum when $V_{gs} = 0$?

2. Analysis of the theoretical transfer characteristic shows that g_m is a function of I_d. Does it matter which value of I_d is used to determine g_m? Why?

Procedure: Part B. Drain conductance and pinch-off voltage

1. With the gate-source voltage V_{gs} set at approximately -0.3 V, measure and record the variation of I_d against V_{ds} from $V_{ds} = 0$ V to 10 V. Use small steps in V_{ds} at low values for V_{ds}.

2. Plot these results and measure the drain conductance $g_d = dI_d/dV_{ds}$ and the corresponding drain "resistance" $R_d = dV_{ds}/dI_d$.

3. At what value of V_{ds} does the normal operating region start (i.e. the pinch off voltage)?

4. Compare this pinch-off voltage with V_{gsoff} from Part A and comment on what these two voltages actually signify.

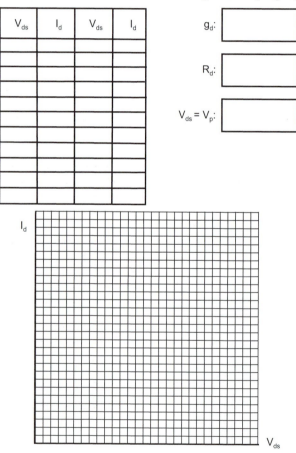

14.13 Common source amplifier

The extremely high input impedance of an FET makes it ideal for a **pre-amp**.
That is, an amplifier which has a modest gain and whose output feeds into the
input of a power amp.
We are going to design an amplifier
with an AC gain of about 20 which
can operate down to about 600 Hz.
The circuit is shown at the right.
When choosing the bias voltage V_{gs},
we must ensure that it corresponds to
the linear region of the characteristic
curve otherwise the output may
become distorted.

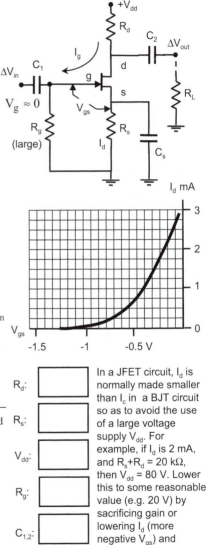

1. The transfer characteristic for the
   2N5484 FET is shown at the right.
   Choose a region of the
   characteristic where the slope is
   approximately linear and estimate
   values of V_{gs} and I_d for a point in
   the centre of this "linear" region.

   V_{gs}: [＿＿＿] I_d: [＿＿＿]

2. Determine the transconductance g_m

   g_m: [＿＿＿]

3. Assume that g_d is very small, say
   40µS, and calculate a value for R_d R_d: [＿＿＿]
   using:
   (use gain $A_v = -20$ $$A_v = -\frac{g_m R_d}{1 + g_d R_d}$$ R_s: [＿＿＿]
   in this formula)
                          $$\frac{-V_{gs}}{I_d}$$
4. Calculate R_s from: $R_s = \dfrac{-V_{gs}}{I_d}$ V_{dd}: [＿＿＿]

5. Let $V_{ds} = V_{dd}/2$ and thus calculate
   V_{dd} from: $\dfrac{V_{dd}}{2} = I_d R_d - V_{gs}$ R_g: [＿＿＿]

6. Choose a suitable value for R_g $C_{1,2}$: [＿＿＿]

7. Choose C_1 and C_2 so that $R_g \omega C = 1$ C_s: [＿＿＿]
   and C_s so that $1/\omega C_s \ll R_s$

In a JFET circuit, I_d is
normally made smaller
than I_c in a BJT circuit
so as to avoid the use
of a large voltage
supply V_{dd}. For
example, if I_d is 2 mA,
and $R_s + R_d = 20$ kΩ,
then $V_{dd} = 80$ V. Lower
this to some reasonable
value (e.g. 20 V) by
sacrificing gain or
lowering I_d (more
negative V_{gs}) and
accepting more
distortion.

Procedure:

1. Construct the amplifier circuit and
measure DC bias conditions, AC
voltage gain and frequency response.
Compare with calculated values.
Comment on any interesting
observations (such as phase
differences, distortions, clipping etc).

2N5484

g   s   d

Bottom view

	Calculated	Measured
V_{gs}		
I_d		
R_g		
V_{ds}		
R_d		
R_s		
V_{dd}		
A_v		
f_o		

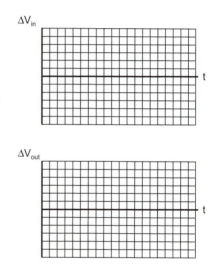

ΔV_{in}

t

ΔV_{out}

t

Questions:

1. Did you have to recompute a value of V_{dd}? If so, why did you select the
value you did and what effect did this have on the expected voltage
gain?

2. Explain why the voltage measured from V_g to earth is about 0 V.

14.14 Logic gates

Basically there are two types of electrical signals: analogue and digital. The foundation of all digital circuits are the logic levels 0 and 1, represented electrically by voltage levels 0 V and 5 V (or sometimes 0 V and −5 V). In this experiment, we introduce some basic digital integrated circuits and how they might be used to construct a digital circuit.

Logic gates are used to represent logic elements of a logic circuit. A logic circuit employs **Boolean algebra** to implement the desired operation. The most useful Boolean expression is perhaps De Morgan's theorem:

$$\overline{(A + B)} = \overline{A} \cdot \overline{B}$$
$$\overline{(A \cdot B)} = \overline{A} + \overline{B}$$

Logic gates are usually supplied on an IC as a series of four on the one chip.

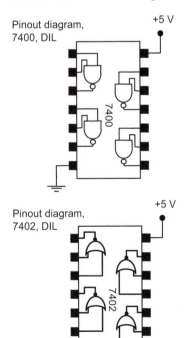

Pinout diagram, 7400, DIL   +5 V

Pinout diagram, 7402, DIL   +5 V

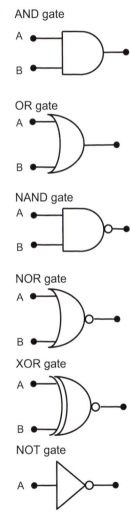

AND gate

A

B

OR gate

A

B

NAND gate

A

B

NOR gate

A

B

XOR gate

A

B

NOT gate

A

Procedure: Part A. Logic gates

1. Using only 2-input NAND gates, construct a logic circuit that operates as a 2-input OR gate. (Use an LED in series with a 200 Ω resistor to indicate the output state of the circuit).

A	B	Out
0	0	
0	1	
1	0	
1	1	

2. Using only 2-input NOR gates, construct a circuit that operates as a 2-input NAND gate.

A	B	Out
0	0	
0	1	
1	0	
1	1	

3. In the circuit below, the output controls a light which is to change state (off to on, or on to off) whenever any one of the three switches changes state. When all switches are earthed, the light is to be off. Construct a truth table for this operation and design and build a logic circuit using 2 or 3-input gates.

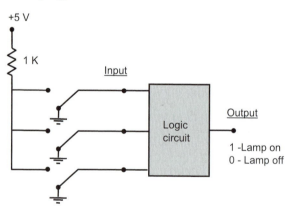

Procedure: Part B. Logic gate characteristics

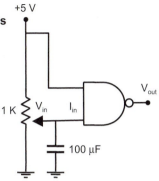

1. Construct the circuit shown with a
   TTL NAND gate and set V_{in} to 0 V.

2. Gradually increase V_{in} until V_{out} goes
   low. Measure the input current I_{in}, the
   output voltage V_{out} and V_{in}.

3. Repeat the above measurements with
   a CMOS NAND gate. Tabulate all
   results.

	V_{in}	I_{in}	V_{out}
TTL			
CMOS			

4. Now construct the circuit so
   as to measure the output
   voltage V_{out} and current I_{out}
   for both the TTL and CMOS
   gates as shown.

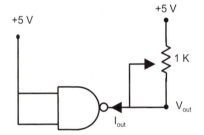

5. With both inputs at +5 V,
   determine the current in mA at
   which the output voltage is
   lifted above the maximum
   value for a low output (0.4 V).

6. With both inputs tied low (to 0 V),
   determine the output current in
   μA required to sustain the output
   high (> 2.4 V).

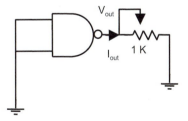

7. Compare your results to those
   which you would expect for
   each type of gate
   construction.

14.15 Logic circuits

Logic circuits are an art. However, there is a systematic way in which a complicated logical function can be designed using the minimum of gates. In this experiment, we use the **Karnaugh mapping** technique to minimise a logic circuit.

1. Draw up truth table and circle min terms

A	B	C	Out	Min terms
0	0	0	0	$\overline{A} \cdot \overline{B} \cdot \overline{C}$
0	0	1	1	$\overline{A} \cdot \overline{B} \cdot C$
0	1	0	0	$\overline{A} \cdot B \cdot \overline{C}$
0	1	1	1	$\overline{A} \cdot B \cdot C$
1	0	0	0	$A \cdot \overline{B} \cdot \overline{C}$
1	0	1	1	$A \cdot \overline{B} \cdot C$
1	1	0	1	$A \cdot B \cdot \overline{C}$
1	1	1	1	$A \cdot B \cdot C$

Pin connections of some common logic gate chips.

2. Construct map as follows:

	C	$\overline{C}$
AB	1	1
$A\overline{B}$	1	
$\overline{A}B$	1	
$\overline{A}\overline{B}$	1	

In a Karnaugh map;
Groups can only be of 1, 2, 4, 8, 16 etc.
The larger the group, the better.
Diagonal groups are not allowed.
Groups can wrap around from edge to edge.
The four corners can form one group.
Don't care states are marked with an X and may be grouped if desired.

(a) Arrange rows and columns with every combination of input, changing only one variable at a time.

(b) Put 1's in boxes corresponding to circled min terms

(c) Draw boxes around groups of 1's. Boxes can only go vertically and horizontally. Boxes can also wrap around. Can only box even groups (powers of 2). Boxes of 3, 5 and 6 etc are not permitted.

(d) Group contents of boxes by ANDs and join together with OR's

Output = $\overline{A}\overline{B}C \, \overline{A}\cancel{B}C \, \cancel{A}\cancel{B}C \, \cancel{A}BC + AB\overline{C} \, AB\overline{C}$

$= C + A \cdot B$

Procedure:

1. A **comparator** tests the value of a pair of two-bit binary numbers A and B. It returns a logic 1 if A is greater than B and a low if otherwise. Complete the truth table for all possible values of A and B and the expected output Q_{exp}

A		B			
A_1	A_0	B_1	B_0	Q_{exp}	Q_m
0	0	0	0		
0	0	0	1		
0	0	1	0		
0	0	1	1		
0	1	0	0		
0	1	1	0		
0	1	1	0		
0	1	1	1		
1	0				
1	1	1	1		

2. Construct a Karnaugh Map from the truth table and derive a Boolean expression which implements the desired logic function of the comparator.

	$B_1 B_0$	$\overline{B_1} B_0$	$\overline{B_1}\,\overline{B_0}$	$B_1 \overline{B_0}$
$A_1 A_0$				
$\overline{A_1} A_0$				
$\overline{A_1}\,\overline{A_0}$				
$A_1 \overline{A_0}$				

3. Draw a logic diagram to produce the Boolean expression. You should be able to do this using 2 inverters, 4 AND gates, and 4 OR gates.

4. Construct the circuit and test its operation using a resistor and an LED on the output. Enter the measured output Q_m in the table above.

14.16 Counters and flip-flops

A bistable **multivibrator** circuit is stable in two states which are called "Set" and "Reset". The stability of such a circuit element is the basis behind digital memory and counters, registers and other sequential control logic circuits. In this experiment, we see how a flip-flop can be used as a binary counter.

The most popular flip-flop configuration is the J-K type which contains "master" and "slave" RS flip-flops in series (although the flip-flop itself is shown here as a single functional block).

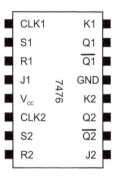

CLK1	K1
S1	Q1
R1	$\overline{Q1}$
J1	GND
V_{cc}	K2
CLK2	Q2
S2	$\overline{Q2}$
R2	J2

7476

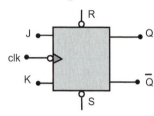

The **asynchronous** inputs S and R are used to set (Q = 1) or reset (Q = 0) the outputs and they override the action table for the J-K inputs. Note, R and S are active low and are thus normally kept high during normal operation. Setting S to 0 sets the Q output (keeping R high) and setting R to 0 with S high resets the flip-flop irrespective of the signals on J and K.

Action table:

R	S	Q	$\overline{Q}$
1	1	no change	
1	0	1	0
0	1	0	1

J	K	Clock pulse 1 to 0
0	0	no change, $Q_{n+1}=Q_n$
0	1	$Q_{n+1} = 0$ (RESET)
1	0	$Q_{n+1} = 1$ (SET)
1	1	$Q_{n+1}= \overline{Q}_n$ toggle

If the inputs of a JK flip-flop are held at 1, then the flip-flop is placed in "Toggle mode".

Clock signal toggles the output.

JK flip-flop as a D-type latch: Clock signal transfers data from D to Q.

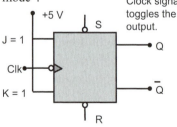

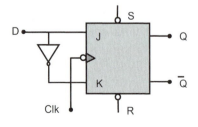

Procedure:

1. The cascaded flip-flop circuit shown can be used as a **counter**. The type of connection shown here produces a ripple (or **asynchronous**) counter. Construct the circuit using a 7476 dual chip and fill in the timing diagram.

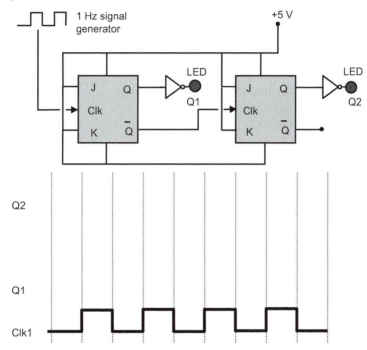

2. Set up a JK flip flop in toggle mode and apply a 100 kHz signal to the Clk input. Display the Clk input and the Q output on an oscilloscope. Measure the time from the trailing edge of the clock pulse to the rising edge of the Q output pulse. Compare with book value.

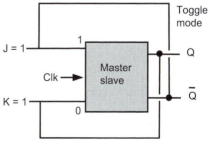

14.17 Op-amps

Having an amplifier in a single package makes it easy to build a variety of circuits. In this experiment, we build an inverting and non-inverting amplifier using an **op-amp**. These two amplifiers form the basis of larger instrumentation and audio amplifiers.

An operational amplifier is a difference amplifier with a very high open loop gain. The connection of external components enables the device to act as an **inverting** or **non-inverting** amplifier. The basic configuration of the op-amp is as follows:

A popular general purpose op-amp is the 741

$R_{in} = 2\ M\Omega$
$R_{out} = 75\ \Omega$
$A_o = 2\times10^5$ (DC)
bandwidth 1 MHz

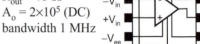

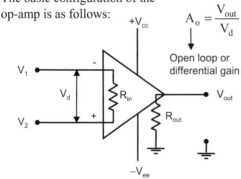

$$A_o = \frac{V_{out}}{V_d}$$

Open loop or differential gain

Connection of **feedback resistors** permit the construction of:

(a) inverting amplifier   (b) non-inverting amplifier

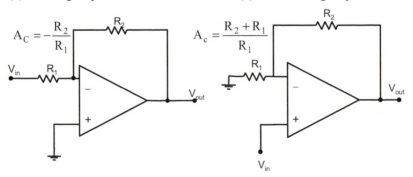

$A_C = -\dfrac{R_2}{R_1}$

$A_c = \dfrac{R_2 + R_1}{R_1}$

Procedure:

1. Design and construct an inverting amplifier which will convert a 50 mV peak-to-peak 1 kHz sine wave into a 2 V peak-to-peak sine wave. Use V_{cc} +10, −10 V.

2. Measure the range of peak-to-peak amplitudes of input signal for which the voltage gain is a constant (i.e. the output voltage is linearly related to the amplitude of the input signal).

ΔV_{in}	50 mV	100 mV	200 mV	300 mV	400 mV
ΔV_{out}					
Gain					

3. Design and construct a non-inverting amplifier which will convert a 100 mV pk-pk 1 kHz sine wave into a 4 V pk-pk sine wave. (Use $R_1 = 1$ kΩ) Compare the measured gain of the circuit with that expected.

4. Measure the gain of the non-inverting amplifier circuit with 3 different values of feedback resistor as shown in the table. (Change the amplitude of the input voltage if necessary).

R_2	ΔV_{in}	ΔV_{out}	Gain	β	Expected Gain

14.18 Comparator

The basic ingredient of an op-amp is a high gain **differential amplifier**. This high gain feature of an op-amp allows it to be used as a comparison detection circuit. A comparator compares two voltages and provides an output (either $+V_{cc}$ or $-V_{ee}$) depending on the relative difference between the two inputs.

Comparator circuits:

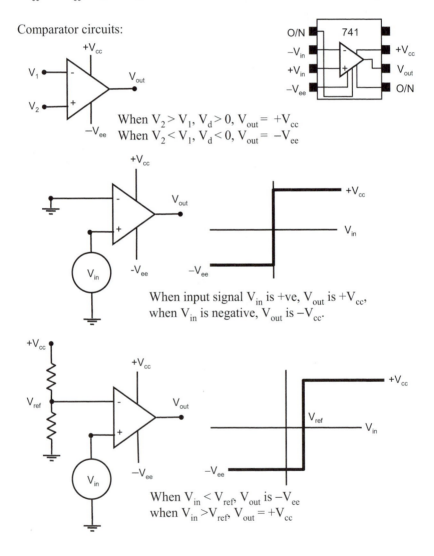

When $V_2 > V_1$, $V_d > 0$, $V_{out} = +V_{cc}$
When $V_2 < V_1$, $V_d < 0$, $V_{out} = -V_{ee}$

When input signal V_{in} is +ve, V_{out} is $+V_{cc}$, when V_{in} is negative, V_{out} is $-V_{cc}$.

When $V_{in} < V_{ref}$, V_{out} is $-V_{ee}$
when $V_{in} > V_{ref}$, $V_{out} = +V_{cc}$

Procedure:

1. Design a circuit using a transistor (as a switch) and an op-amp which will turn on an LED when a resistor R with a value less than 1 kΩ is connected across the input. If no resistor is connected, or the resistance is greater than 1 kΩ, then the LED is to be off.

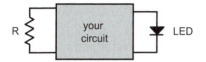

Draw your circuit and explain its operation.

2. Construct the Schmitt trigger circuit shown and measure the DC input voltages at which the output changes state (Use $V_{cc} = +12, -12V$). Compare with expected voltages.

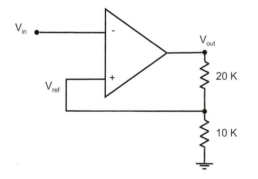

14.19 Integrator

Why is an integrator useful? Many physical measurements involve the integration of a quantity with respect to time. An integrating circuit gives an output whose amplitude is the integration of the input voltage w.r.t. time. The circuit does this integration instantly and does not care how complicated the input signal is. Many electrical measurements involve the integrated output and such circuits are very common in scientific instrumentation.

Integrator

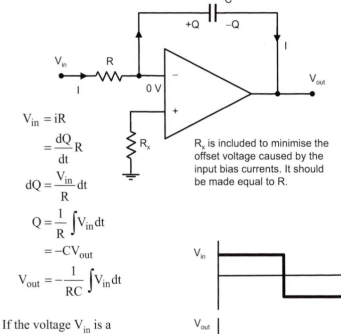

$$V_{in} = iR$$

$$= \frac{dQ}{dt} R$$

$$dQ = \frac{V_{in}}{R} dt$$

$$Q = \frac{1}{R} \int V_{in} dt$$

$$= -CV_{out}$$

$$V_{out} = -\frac{1}{RC} \int V_{in} dt$$

R_x is included to minimise the offset voltage caused by the input bias currents. It should be made equal to R.

If the voltage V_{in} is a constant, then the output is simply:

$$V_{out} = -\frac{V_{in}}{RC} t$$

Procedure:

1. We wish to convert a 1 V pk-pk square wave to a 2 V pk-pk triangle wave. Using a 741 op-amp, design such a circuit. (Let $R_x = R$)

2. Construct the circuit but connect both inputs to earth (0 V). Connect a voltmeter to the output and determine whether or not the output changes with time (drift). Measure the drift (V_{out} per second).

If there is drift, then connect a 1 or 2 MΩ resistor across the capacitor. If the output of the circuit is "stuck" on +15 or -15, then discharge the capacitor with a piece of wire before taking measurements of drift.

> How does the 1 MΩ resistor reduce drift?

3. Disconnect inputs from 0 V and then connect to square wave oscillator. Make sure that the pk-pk input voltage is exactly 1 V and tune the circuit to obtain exactly a 2 V pk-pk output signal.

4. Test the circuit at increasing frequencies and record and comment on your observations.

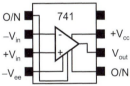

O/N 741

$-V_{in}$ −

$+V_{in}$ +

$-V_{ee}$

$+V_{cc}$

V_{out}

O/N

Component values

Resistors

Resistors used in electronic circuits are colour-coded to indicate their resistance (in ohms) and the tolerance. The tolerance indicates by how much the actual resistance of the device may differ from its nominal value. Usually, a tolerance of 1 to 5% is acceptable.

Colour	1st band 1st figure	2nd band 2nd figure	3rd band multiplier	4th band tolerance
Black	0		1	
Brown	1	1	10	1%
Red	2	2	10^2	2%
Orange	3	3	10^3	
Yellow	4	4	10^4	
Green	5	5	10^5	
Blue	6	6	10^6	
Violet	7	7	10^7	
Grey	8	8	10^8	
White	9	9	10^9	
Silver			10^{-2}	10%
Gold			10^{-1}	5%

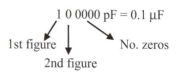

e.g., blue green violet = 650 MΩ 5% tolerance

Resistors are also manufactured to a maximum power dissipation rating. However, the power rating is usually not specified on the resistor itself, but a good estimate may be made from the physical size of the component. Most resistors in electronic circuits are 0.25 to 1 W power rating.

Capacitor markings:

1st figure
2nd figure
number of zeros (for pF)
Letter: tolerance

Example: 104 K

1 0 0000 pF = 0.1 μF

1st figure   No. zeros
2nd figure

15. Index

Index